Peregrine
Falcon
Prairie
Falcon
Merlin
American
Kestrel
White-tailed
Kite
Swallow-tailed
Kite
Northern
Harrier
Mississippi
Kite
Turkey
Vulture
Golden
Eagle
Bald Eagle

HAWKS IN FLIGHT

For Bob Wallis, with best birthday wishes —
Scott Weidensaul

Dear Bob,
Thanks for a marvelous week in a magical place.
[signature] Johnson

Bob -
I can't imagine a more magical place to celebrate a birthday!
Hope you have many great memories!
Sara Morris

To - BoB
Happy 71 ST. B-Day.
Your Friend,
Ji[signature]

HAWKS IN FLIGHT

SECOND EDITION

The Flight Identification of North American Raptors

Pete Dunne, David Sibley, and Clay Sutton

Houghton Mifflin Harcourt
Boston New York 2012

www.hmhbooks.com

Library of Congress Cataloging-in-Publication Data is available.

ISBN 978-0-395-70959-7

Printed in China
Book design by Eugenie S. Delaney

SCP 10 9 8 7 6 5 4 3

PREVIOUS SPREAD: The Gray Ghost. Among frequently seen hawks, perhaps no other raptor commands the esteem among hawk watchers as the adult male Northern Harrier. JL

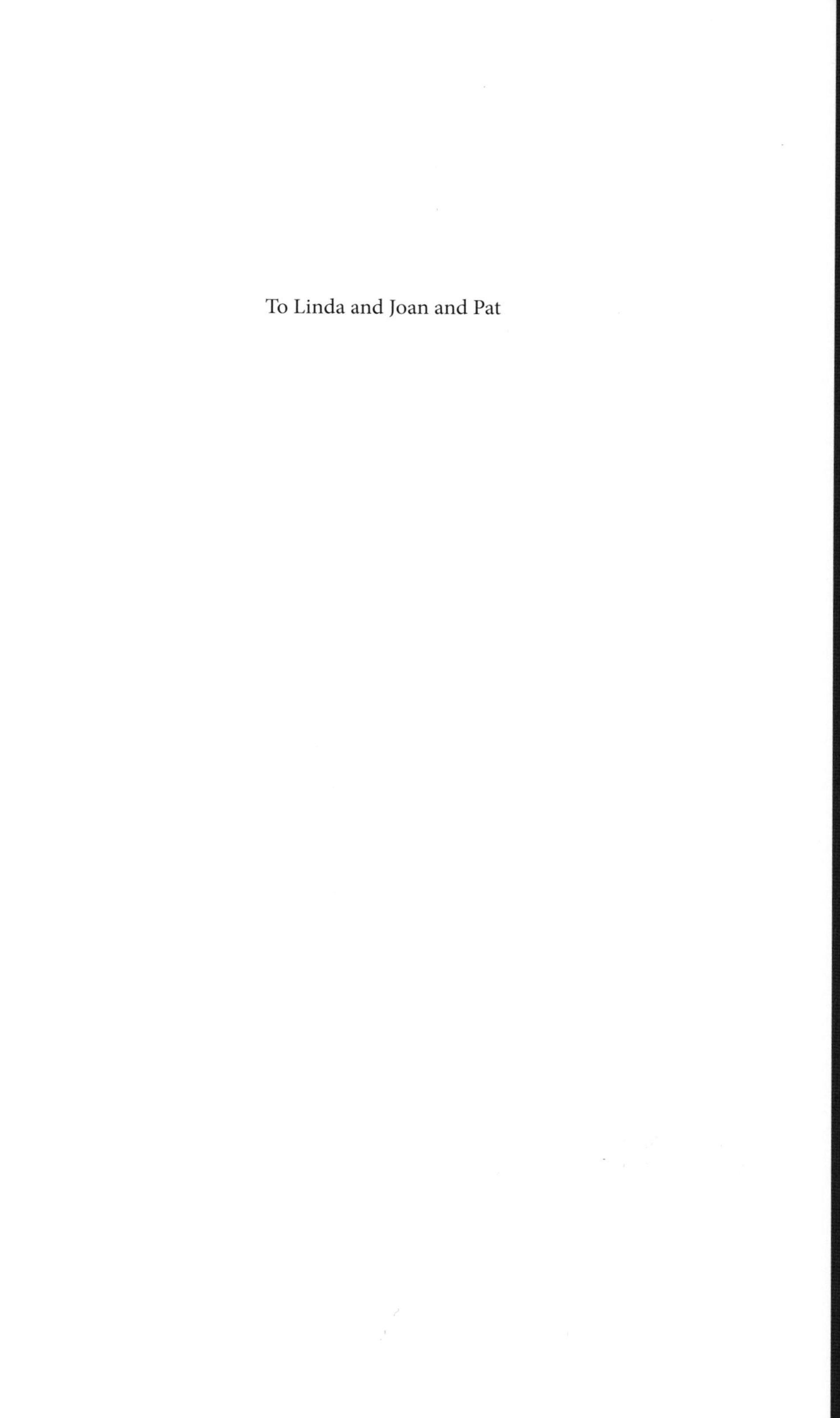

To Linda and Joan and Pat

Contents

LEFT: The Prairie Wraith. There are many reasons to watch hawks in the American West, and the fleet Prairie Falcon is one of the best. JL

Acknowledgments

If the authors were indebted to hundreds of friends, colleagues, and fellow hawk watchers for their contributions to the first edition of this book, that number must now range in the thousands. Raptor identification is a challenge that has brought a host of keen minds to bear. In honesty and in fairness, all individuals who find parts of their identity in the label "hawk watcher" deserve recognition in these pages because all have made their special contribution.

There are, however, some individuals whose contributions to this work deserve special recognition, starting with those whose counsel figured so prominently in that first edition of *Hawks in Flight.* They are: Harold Axtel, Pete Bacinski, Seth Benz, Rick Blom, Pete Both, Jim Brett, Bill Clark, Harry Darrow, Bob Dittrick, Gladys Donohue, Jim Dowdell, Howard Drinkwater, Mary Jane Evans, Fred Hamer, Greg Hanisek, Michael Heller, Steve Hoffman, Paul Kerlinger, Harry LeGrand, Arnie Moorhouse, Al Nicholson, Frank Nicoletti, Sam Orr, Roger Tory Peterson, Richard Porter, Steve Potts, Noble Proctor, Chandler Robbins, Jim Ruos, Will Russell, Fred Sibley, Gerry Smith, Ted Swem, Fred Tilly, Hal Wierenga, and Floyd Wolfarth. Most of these individuals still enjoy what are, for the most part, ventral views of hawks. A few, owing to a change in watch sites, are now specializing in dorsal views. All enjoy our thanks and appreciation.

This second edition was aided by the helpful suggestions of several individuals, most notably William S. Clark, whose critique of our first work was the basis for changes relating to description and terminology.

The assistance of three individuals, in particular, figures prominently in this book. These individuals are Jerry Liguori, Frank Nicoletti, and Allen Fish, director of the Golden Gate Raptor Observatory in San Francisco. All gave freely of their time and their considerable expertise at the onset of this project. In deference to their contribution, when referring to this book please add their names to ours. *Hawks in Flight* by Dunne, Sibley, and Sutton (and Fish, Liguori, and Nicoletti).

Other individuals offered their counsel and wisdom both before and during the writing of this book. They include Jorge Montejo Diaz, Miriam Moore, John Schmidt, Sheri Williamson, Tom Wood, Laurie Goodrich, Louise Zemaitis, Michael O'Brien, Charles Duncan, Jeff Bouton, Jeff Dodge, Sandy Sherman, Fred Mears, Joel Simon, and Julio Gallardo.

While working on the manuscript, we found Jerry Liguori's *Hawks from Every Angle,* Brian K. Wheeler and William S. Clark's *A Photographic Guide to North American Raptors,* and Wheeler's *Raptors of Eastern North America* and *Raptors of Western North America* immensely helpful. Other resources include pertinent species accounts in the Birds of North America series, and a wonderfully crafted, privately published guide to the raptors of southern Texas written and illustrated by the late Gladys Donahue. Our hats are off to her and her pioneering efforts.

Photos for this edition were garnered from a number of very fine photographers, and their inclusion in these pages greatly enhances this book, as you are destined to discover. Principal photographers were Jerry Liguori, Ned Harris, Kevin T. Karlson, and Clay Sutton. William S. Clark, Don Freiday, Tom Johnson, Tony

Leukering, Jim Lish, Michael O'Brien, Luke Ormond, Linda Dunne, Ted Swem, and Merrill Cottrell provided key images that ably filled many crucial gaps.

Kevin Karlson and Pat Sutton were particularly helpful in the gathering and processing of images.

This revision has been a protracted affair, tabled and picked up multiple times as the authors dealt with assorted other writing projects. We owe a debt of thanks to our editor, Lisa White, who demonstrated uncommon patience (not to mention faith); and to our agent, Russ Galen, whose unwavering encouragement was equaled only by his diplomatic skill.

Finally, and almost without saying, the authors affirm their individual and collective gratitude to the three individuals whose patience, indulgence, and support are the very foundation of all we do. Their names are Linda Dunne, Joan Walsh, and Pat Sutton, and they are, to our great fortune and their periodic despair, related to the authors by marriage.

Thank you for everything. Sorry about the lost weekends. We promise not to tackle another edition for at least a decade. Probably two.

Introduction

Some 24 years and 48 migratory periods have passed since the publication of *Hawks in Flight* — a book intended to impart to readers the information and skills needed to identify distant birds of prey. To our astonishment, more than 60,000 copies of that first edition were purchased by both ardent and incipient hawk watchers. Authors and publishers could hardly ask for more.

Readers are different. Almost before the ink had dried on the pages, hawk watchers were clamoring for an edition of *Hawks in Flight* that would include those species not covered in the first edition — raptors that have limited distributions in North America and whose seasonal redistributions are limited; raptors that we, the authors, chose not to include among 23 species of hawks that enjoy widespread distribution and migrate great distances.

Yes, readers understood our rationale, appreciating the fact that reducing variables is one of the fundamental tenets of hawk craft. But they, and finally we, understood that, in our efforts to serve probability, we had shortchanged hawk watchers in the higher currency of *possibility* and failed to serve those living in or visiting raptor-rich southern and border regions.

This second edition, therefore, includes *all* of the birds of prey that have established breeding populations in the United States and Canada. Wholly revised, integrating the best elements of the first edition with fresh insights, this new edition nevertheless remains true to the principle of simplification. The 11 new species incorporated into this second edition — California Condor, Hook-billed Kite, Snail Kite, Common Black-Hawk, Harris's Hawk, Gray Hawk, Zone-tailed Hawk, Short-tailed Hawk, White-tailed Hawk, Northern Caracara, and Aplomado Falcon — are spatially segregated in the book just as they are separated from most hawk-watching sites in geographic fact.

When you visit areas where borderland buteos like Gray Hawk or Common Black-Hawk are found, or Florida where Short-tailed Hawk and Snail Kite occur, you can also find them in this book. But if you are hawk watching where the chances of seeing these regional specialties are remote, the identification challenge has been simplified by excluding these unlikely species from the ranks of likely ones.

Speaking of remote possibilities, several species of raptors whose appearance in North America is accidental (e.g., Eurasian Kestrel and Red-footed Falcon) are *not* included. Our efforts to serve possibility did not go so far as to embrace improbability. If you care to become familiar with Old World raptors, we recommend you try

LEFT: The Black Warrior. John James Audubon may have officially named it the Harlan's Hawk, but his praise for the bird is best evident in his sobriquet. JL

Flight Identification of European Raptors and *The Raptors of Europe and the Middle East* — guides whose focus begins where this one's ends. For borderland possibilities such as Roadside Hawk and Crane Hawk, we recommend Steve N. G. Howell and Sophie Webb's *A Guide to the Birds of Mexico and Northern Central America.*

The original *Hawks in Flight* was crafted to do two things: First, to impart to readers the hints and clues that distinguish one distant bird from another — from plumage characteristics to the subtleties relating to shape, wingbeat, and manner of flight. Second, to convey to readers the process of seeing distant birds critically and building confident identifications from the evidence amassed.

Most field guides tell users what to look for to distinguish one species from another. Our broader ambition was to tell users how to look for those distinguishing characteristics and how to process and interpret that information once they got it. This second edition follows in the footsteps (more nearly the wingbeats) of its predecessor.

Must you understand how or why the book was written and organized to benefit from it? No. Be assured that the text, photos, and drawings are orchestrated to meet your needs and the challenges of hawk watching.

For example, the bulk of the species in this book have been grouped with an eye toward shared traits and similarity in behavior or appearance. A taxonomist might cringe at the thought of lumping eagles and vultures together, but any hawk watcher will understand the rationale. At a distance, eagles and vultures look much alike.

In this same vein, the drawings are black and white, not color. While color is, at times and under the right conditions, a wonderful aid to identifying some raptors, too often birds of prey are identified at distances or under light conditions that make color difficult or impossible to determine. But in the interest of covering all bases — and unlike in the first edition — we have included color photographs in this second edition. This decision was based partly upon the sheer wealth and quality of the color images available today, and partly on the fact that, thanks to more than 20 years of optical refinement, hawk watchers today are able to perceive more color and detail than they could when that first edition saw the light of day.

The captions accompanying the illustrations are concise; they were written to facilitate quick referral or to reinforce key distinguishing traits. The body of the text was written to be both informative and engaging. Descriptions and material relating to species and groups of species are imparted in a systematic fashion. The basic format includes three elements:

1. A description of each species, its range, and its behavior, so readers will understand something of the special nature of each bird.
2. A detailed account of the identifying field marks, starting with fundamental and familiar field marks relating to plumage and shape, and concluding with subtle and subjective elements relating to movement and flight.
3. A summary section that compares and contrasts species of similar appearance, a process that approximates the challenge hawk watchers face in the field.

Much thought went into selecting words or analogies that would fix subjective images in readers' minds. Humor and, at times, an irreverent style are integral to the text. We hope you find this book engaging and we hope that your hawk-watching skills climb to levels beyond the reach of your dreams — even beyond the reach of this book! Somewhere, on celebrated ridge tops, in strategic city parks, and on isolated fingers of land projecting into lakes and oceans, individuals are gathering the insights that will vault raptor identification to even higher levels of perfection — experts who are destined to one day codify their knowledge in a book like this.

Maybe one of these people is you. We hope this book serves as a valuable point of departure, no matter where your interest in raptors takes you or how high that interest soars.

The Flight Identification of Raptors

From the Shotgun to the Sublime

There is nothing mystical about identifying distant birds of prey. The skills needed to pin a name to a bird that might be little more than a speck on the horizon are firmly rooted in the identification system pioneered by Roger Tory Peterson and immortalized in his famous field guide. Part discipline, part deductive process, hawk identification demands that birders not rely on just one or two distinguishing characteristics to make an identification. Indeed, hawk watchers must often combine a number of hints and clues in order to craft an identification that conforms to the bird in the sky.

Some clues may be blatant, even familiar — like the crooked-wing configuration of an Osprey or the single broad white band that bisects the tail of a high-flying adult Broad-winged Hawk.

You say that your field guide shows two bands, even three? We know. That's why you are having so much trouble identifying distant birds of prey. Most field guides depict birds as observers might see them at close range or even perched. On distant, high-flying Broad-wingeds it is common for only a single wide white tail band to be visible.

Some of the clues are subtle, even subjective — like the fluid, undulating wingbeat of a Peregrine or the way American Kestrels seem to float in a glide (while Merlins sink), or the way high-flying Sharp-shinned Hawks appear headless — an accipiter that is all wings and tail.

You say your basic field guide doesn't mention these things? We know. That's why we wrote this book.

Go to one of the hundreds of hawk-watching junctions scattered across North America. Watch a skilled hawk watcher pit his skill against the horizon and you'll see names pinned to birds at distances once deemed impossible. It won't be the first time such advances in bird identification have been made.

LEFT: Hawk watching at its finest: the "River of Raptors" — an enormous kettle of hawks — circles over Cardel, in Veracruz, Mexico, in early October. This group is primarily Broad-winged Hawks, with a few Swainson's Hawks and Turkey Vultures mixed in. Can you spot the single Peregrine? CS

Consider: At one time, the limit of bird identification skills was defined by the effective range of a shotgun. A little more than a century ago, an ornithologist seeking the identity of a bird needed to collect it.

Birds then, as now, were skittish and uncooperative creatures, suspicious of humans and their motives, and not inclined to tolerate prolonged scrutiny or close approach. Shooting the bird overcame these obstacles. With the bird in hand, ornithologists were able to examine the specimen at length and at leisure. Identifications were made on the basis of traits (such as the color of a bird's eye or the length of its tarsus) that could be determined by studying the bird in the hand. If an ornithologist had second thoughts about his initial identification, he could pull the bird from his collection, reexamine it, and confirm or refute his original conclusion.

This was a very workable, very dependable system. But it had its drawbacks. Most obviously, it was rough on the bird. If the particular problem was identification or documentation, then the shotgun approach worked fine. For the student of bird behavior, however, it offered insight into only one of the facets of a bird's life — its death.

Furthermore, given the limitations of a shotgun's range, it essentially restricted bird study to about forty yards — the effective range of light birdshot.

Around the turn of the twentieth century, scientists began to adopt new tools that helped to close the distance between birds and bird students: field glasses, later replaced by binoculars. These optical devices allowed birds to be studied in the field in real time, and permitted details to be seen beyond the range of the unaided human eye. It soon became apparent, however, that the details visible in the glasses were not necessarily the same details that could be seen in the hand.

Distance transforms things! What might be a pattern of dots around the eye fuses into a solid eye-ring under the alchemy of distance and glass. Pale wing coverts that might differ from surrounding feathers by only a shade in the hand coalesce into a contrasting wing bar in the field.

In 1934, a 25-year-old artist and bird enthusiast published a book that compiled the newfound knowledge concerning bird identification. Its principal feature was black-and-white (and a few color) plates that depicted birds as they might be seen in the field, in side profile, using the finest 4x and 6x optics of the day. Tiny arrows drew attention to those visible traits that served to distinguish one species from the next. The author called them "trademarks of nature." In later editions of his guide, he called them "field marks."

The artist and author was, of course, the late Roger Tory Peterson. And his guide, now in its sixth edition, remains an invaluable birding resource — a Rosetta stone to the birds. But just like the shotgun school, the new approach to bird identification, as codified in the Peterson guide, had drawbacks, too.

For one thing, live birds have considerably greater freedom of movement than dead ones. Looks at birds in the field might be momentary. Identification often has to be made quickly or under unfavorable conditions. Sometimes key field marks cannot be seen because of obstructing foliage, a poor angle, too great a distance, or plumage characteristics that are masked by low light conditions or harsh backlighting. The element of certainty enjoyed by proponents of the shotgun school diminished with this new approach. If the bird flew before observations were complete or if an observer had second thoughts, there was no equivalent of going to the specimen drawer and reconfirming the original identification — the bird was gone.

A subtler problem introduced by the Peterson System was that it tended to lock practitioners into a rigid mindset. It focused on individual traits of individual birds, most commonly traits relating to plumage. To this day, some bird watchers hesitate to make an identification unless birds present a classic, stationary, side profile.

Finally, and perhaps most troublesome, the simple fact was that the field marks used by early field birders to clinch an identification could often be seen only at close range — sometimes, given the conditions, at closer than shooting range.

When those first hawk-watching pioneers clambered onto the ridge tops and raised their glasses from coastal vantage points, they carried with them Roger Peterson's system of identification. Its limitations in the hawk-watching arena immediately became apparent.

On September 17, 1935, Maurice Broun, Hawk Mountain Sanctuary's first curator, recorded 978 Red-shouldered Hawks and 2,175 Broad-winged Hawks from the North Lookout at Hawk Mountain in Pennsylvania. As Maurice noted in his private journal: "I was at my wit's end differentiating Broad-wings from Red-shoulders."

The problem was a good deal less troublesome than Maurice made it. The birds were *all* Broad-winged Hawks, as everybody knows now (and as Maurice, after his initial frustration, also concluded). The field mark he relied upon to distinguish these two similar buteo species was the width of the bands on the tail — a field mark that was easily noted on perched adult birds at close range. Maurice's problem was that he was seeing both immature birds and adults and was forced to make identifications at distances too great to let him see the bands on a tail.

Hawk watching stretched the Peterson System to its limits, and the system fell short. It just didn't work at the distances and the angles that people encountered in the hawk-watch arena.

The new system of hawk identification developed not overnight but gradually, requiring many seasons and involving a number of keen, attentive minds (including Maurice Broun's and Roger Peterson's). In time, field marks that did not work in the hawk-watching arena were weeded out. New ones, tailored to the identification of birds in flight, were discovered. Cooper's Hawks, for example, appeared to have a large head, while the head of the very similar Sharp-shinned Hawk seemed small. It became apparent that Red-shouldered Hawks tended to flap more than Red-tailed Hawks, and they frequently flew on the off-wind side of the ridge — the side away from the updrafts that big burly Red-taileds favored. (Being smaller and lighter, Red-shouldereds find the less-buffeting air on the lee side more to their liking.)

It wasn't that distant birds of prey showed no field marks. They simply showed *different* field marks — distinguishing characteristics that related to shape, behavior, and movement as opposed to (or more nearly, in addition to) traits linked to plumage.

In time, a subtle change in thinking occurred that was, at first, largely overlooked. A bird was identified as a particular species no longer because it *had* or *showed* this or that field mark, but because it *seemed* to have this feature or *tended* to exhibit this particular behavior while another species *tended not* to do so. Identification became more subtle and more subjective, and a greater level of uncertainty became acceptable.

Just because a distinguishing characteristic was not infallible didn't mean it wasn't a useful aid to identification. It *was* valuable! It just wasn't definitive; it wasn't a trademark, and it wouldn't stand alone. To mitigate the uncertainty imposed by distance, hawk watchers responded by compounding the number of distinguishing characteristics needed to gain a measure of confidence — hints and clues relating to behavioral traits; the rhythm and cadence of a bird's flight; its overall color, shape, and size; and plumage characteristics that vaulted the distance (and indeed, some characteristics that became apparent only with *added* distance). Taken in sum, these compounded hints and clues build a case that leads to an identification, or at least a hypothesis to test: "I think the bird is this. Now, on the basis of what I continue to see, is it?"

Is every effort to identify distant birds of prey so complicated? Certainly not. Some identifications are easy because some characteristics are obvious. In many instances, those who are used to using classic field marks to identify birds will find themselves on familiar footing as they edge out onto the hawk watcher's plateau. The reddish tail brandished by most adult Red-tailed Hawks is as prominent when the bird is soaring as when it is perched. Given good light conditions, the white head and white tail of an adult Bald Eagle are visible (and diagnostic) whether the bird is 100 or 5,000 yards away.

But certain aspects of raptor identification will take some getting used to. One example is the reduced emphasis on plumage. For a number of species, particularly buteos, plumage plays a key role in identification (as well as determining age, sex, or subspecies). But for most distant raptors, plumage characteristics are not as valuable (or even discernible) as characteristics relating to size, overall shape, and manner of flight.

Color, a key element in the identification of warblers and waterfowl, has diminished significance in the hawk-watching arena. Often hawks are flying at distances or under light conditions that make it impossible to note anything but blatant patterns or a bird's overall color (the homogenization of plumage subtleties). Second- and third-year Bald Eagles, for example, are a mottled mix of pale and dark brown feathers. At a distance, the mottling disappears, homogenizing into an overall mocha- or latte-colored bird.

Possibly we should retract, or at least qualify, our initial assertion that there is nothing mystical about hawk identification. If magic is defined as the art of illusion, then the magic inherent in the wind, in the angle of the sun, and in the contours of the earth below can change the shapes of birds so that they look very different from the way they appear in the hand — so that they even look, at times, like entirely different species.

A Red-tailed Hawk turning lazy circles in a rising thermal is the very picture of a buteo: wings fully extended, tail fanned. But when the same bird glides to the next thermal, it draws in its wings, closes its tail, and suddenly becomes more falconlike. Then again, riding the updraft off a ridge in a 50-knot wind with wings folded flat against the body, a Red-tailed takes on the characteristics of "a flying cinder block," as the late Floyd P. Wolfarth was fond of describing the bird. It takes a measure of magic to work such shape-shifting transformations.

The element of distance alone is enough to transform one bird into another. The bird in the bush looks different from the bird in the hand. Individual feathers merge into breast streaks and wing bars. Colors fade but patterns sharpen. When the bird leaves the bush and enters the ozone, even bodily features — head, tail, and wings — blur and blend into a composite image, a generalized shape.

In accipiters, for example, wing shape, to a large extent, accounts for the large-headed appearance of a Cooper's Hawk and the small-headed look of a Sharp-shinned Hawk. A soaring Cooper's Hawk's wing has a straight leading edge and the bird's head projects well beyond it. Even at high altitudes, the head remains discernible. The wing of a soaring Sharp-shinned Hawk, on the other hand, juts forward at the wrist. As distance increases, the distinction between head and wing becomes indiscernible. The Sharp-shinned's head gets lost in the gully between the wings. The result? At a distance, a Sharp-shinned appears headless, a flying mallet (with the broad, short wings serving as the mallet's head, and the body and tail becoming the handle). A Cooper's Hawk resembles a flying cross or crucifix.

The wings of a Peregrine Falcon are undeniably long and thin — eminently falconlike. But in a full soar, a Peregrine fans its tail so that the outer tail feathers nearly touch the trailing edge of the wing. The astounding breadth of the tail effectively masks the length of the wing. In general impression and shape, a distant soaring Peregrine is very reminiscent of a soaring Broad-winged Hawk — a buteo and a most unlikely candidate for confusion.

Understand: Raptor identification is little concerned with actual size or proportions. It is concerned with how birds *appear* at a distance. Often the two aspects differ. Take, for example, Northern Harrier. Some field guides (including some raptor guides) ascribe dark streaking on the chest to juvenile harriers. Seen closely and well, the birds *are* streaked. But at the distances that birds are commonly seen by hawk watchers, the diffuse streaking melds with the cinnamon underparts of juvenile birds and becomes indiscernible.

Or consider tail length in Red-tailed Hawks. Juvenile Red-tailed Hawks have longer tails than adults, and the juveniles' narrower wings make their tails *appear* even longer. The combination of long narrow wings and longer tail makes juvenile Red-taileds appear rangier than adults.

In sum, the flight identification of distant raptors is a blend of identification skills, deduction, and even a measure of conjecture. The system offered in this book has evolved to meet the challenge presented by birds flying at distances beyond the reach of more static, more traditional identification methods. But just like its predecessors, this system of identification has its drawbacks. Certainty diminishes once again. Anyone who enters the hawk-watching arena must understand that 100 percent positive field identification is an ideal rather than a certainty.

A measure of uncertainty (and humility) is the price hawk watchers pay for doing what would have seemed impossible 100, 50, or even 24 years ago, when the first edition of this book was published. Like Kierkegaard's leap of faith, we give up something (utter certainty) in order to gain so much more (confident identifications at the near limit of detection). From this footing it is possible to gain an understanding of birds and their complex movements from horizon to horizon. Such is the nature of hawk watching and the scope of this book.

Buteos That Migrate

The Wind Masters

SPECIES
Red-tailed Hawk, *Buteo jamaicensis*
Red-shouldered Hawk, *B. lineatus*
Broad-winged Hawk, *B. platypterus*
Swainson's Hawk, *B. swainsoni*
Rough-legged Hawk, *B. lagopus*
Ferruginous Hawk, *B. regalis*

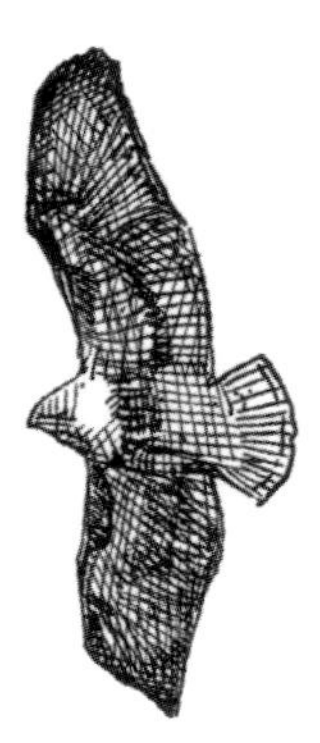

Buteos are a diverse group of medium to large hawks that excel in the art of soaring. Some are forest birds, some are at home in open country. All are wind masters, able to tease lift out of temperature-troubled air and soar for long periods on set wings, using this mode of locomotion for hunting, courtship, and long-distance migration.

Unlike accipiters and falcons, which are united by similar habitats and hunting methods, buteos exhibit great diversity and versatility among species — even to the point of commonality with nonbuteo species, such as falcons and accipiters, whose niche is more specialized. Rough-legged Hawks (a buteo) and Peregrines (a falcon) may use the same nest ledge in alternate years in the Brooks Range of Alaska, and Red-shouldered Hawks (a buteo) will share a wooded hillside with a Northern Goshawk (an accipiter) in New England. A Swainson's Hawk (a buteo) will course over the same short-grass prairie as a Northern Harrier in Colorado in search of newly fledged ground-nesting birds. And whether an incautious ground squirrel will fall prey to a Ferruginous Hawk, a Prairie Falcon, or a Golden Eagle depends upon who gets there first.

Rough-legged, Red-tailed, and Ferruginous hawks can hover-hunt like an American Kestrel. Broad-winged Hawks will snatch dragonflies on the wing in the

ABOVE: Adult Red-tailed (left) with adult Red-shouldered (right). The more compact shape and uniform body color of the Red-shouldered contrasts with the more "muscular" and patterned Red-tailed. LEFT: Dorsal view of an adult Red-shouldered Hawk. Although the proportions and shape are all buteo, the pattern of an adult Red-shouldered is more colorful than that of most buteos. JL

manner of a Mississippi Kite. And Red-shouldered Hawks perch-hunt with a finesse that would do credit to the most accomplished Northern Goshawk (and show more patience as well). In short, if there is a niche to be filled by a predator wearing feathers, chances are that a buteo is firmly entrenched there.

Twelve buteonine species breed in North America north of the Rio Grande. Fully half of these have ranges that limit their occurrence in the United States to extreme southern and southwestern border regions; they are treated in Chapters 10 and 11. Six qualify as moderate- or long-distance migrants and fall within the scope of this chapter.

MIGRATION

When many raptor enthusiasts speak of hawk migration, their thoughts automatically turn, not surprisingly, to buteos. Geographically speaking, buteo migration is more broad-based and more easily witnessed than the more coastal migrations of accipiters and falcons. In addition, several species, most notably Broad-winged and Swainson's hawks, are given to spectacular migratory aggregations that may involve hundreds, thousands, and (at key concentration points) even tens of thousands of birds.

Few people would fail to be impressed by the sight of thousands of Swainson's Hawks rising from the Texas plains in spring, or by a swirling tornado of Broad-winged Hawks crossing the Connecticut River Valley in mid-September. Often as not, the peak raptor flight at Hawk Mountain in Pennsylvania occurs in November, not September, and is dominated by squadrons of Red-tailed Hawks that ride the ridge in groups of four, five, or even ten. In late April, at Whitefish Point, Michigan, the spring surge of Sharp-shinned Hawks can be spectacular. But this surge of smaller hawks is overshadowed (in the minds of many observers) by the ranks of Rough-legged Hawks moving along the Lake Michigan shore.

Buteos migrate in a broad, sweeping wave across the continent unless some outside force deflects them. Such forces are called *diversion lines* and *leading lines*. Leading lines attract birds of prey. Diversion lines impede their passage and deflect them.

A diversion line might take the form of a large body of water, such as one of the Great Lakes, which acts as a barrier to migration. When migrating birds encounter water, they face a choice. They can either cross and expend energy over a thermal-poor and updraft-free substrate or make the more energetically correct choice and alter their course to follow the contours of the shore. Unless the crossing is short or winds particularly favorable, the strategy of most buteos is to hug the shoreline — sometimes right at the water's edge, sometimes a few miles inland where thermals are more abundant. Thus concentrated and directed, these streams of birds pass over such well-known hawk-watch sites as Braddock Bay near Rochester, New York, and Derby Hill, New York, in the spring, and Hawk Cliff, Ontario, and Duluth, Minnesota, in the fall.

The great concentrations of migrating Broad-winged Hawks near Corpus Christi, Texas, are a classic example of a water barrier at work. In the grand scale, the Gulf of Mexico directs these birds en route to Central America west and then south along its coast. At Hazel Bazemore Park in Corpus Christi, the shore angles south, and Corpus Christi Bay juts across the path of the birds, diverting them inland and serving as a geographic incentive for birds to gang up just a little bit more before leaving Texas and entering Mexico.

An approaching cold front is another form of diversion line. Hawks moving north in the spring (and at times and in places in the fall) often meet the edge of a front approaching from the northwest. The poor weather and turbulence associated with these fronts constitute a physical barrier. Birds adjust their course to fly along the leading edge of the front, concentrating along its length, forming a ribbon of birds that advances just ahead of the wall of clouds. At such times, any place that lies ahead of the front becomes a hawk-watching hotspot — a temporary fortune that lasts until the front sweeps overhead.

Mountain ridges also concentrate migrating raptors (particularly buteos), but the mechanism is different. Ridges concentrate birds not by blocking their path but by offering mile upon easy mile of energy-conserving updrafts. As long as the winds strike the ridge so that they are deflected upward, and as long as the ridge runs in the direction hawks want to go, birds will concentrate along the windward side of the crest, getting the energetic most out of a free ride. This is the secret behind the magic of Hawk Mountain, Pennsylvania, where hawk watching was born.

Even without an updraft, ridges can be a boon to migrating raptors early and late in the day. As the sun rises and sets, vertical rock faces absorb heat energy more efficiently than do horizontal terrains. This heat energy is passed into the surrounding air, which rises, causing thermal lift — a rising column of air that hawks can ride.

All migrating raptors use thermals for lift. The thermals are widespread (not limited to ridges) and in fact form anywhere that the ground heats at different rates. For example, a black asphalt parking lot will heat more rapidly than an adjacent, well-vegetated city park. When the air lying over the surface of the lot becomes warmer than surrounding air by as little as one or two degrees, it rises, forming a thermal.

On days when thermal production is optimal and thermals are widespread, hawks prefer thermals to updrafts off a ridge to gain lift. Ridge updrafts are used principally on days of heavy cloud cover (when the lack of sunlight hampers thermal production) and on days when strong winds strike auspicious ridges directly — conditions that cause strong updrafts but disrupt thermal production by rapidly mixing the air.

If a ridge does not lead in the direction migrating buteos want to travel, or if it is not continuous or meanders too much across a landscape, raptors may still be drawn to the updrafts that wind and geography conspire to create. They use these opportune updrafts as if they were thermals. Birds will circle in an updraft, riding it upward. When they've gained the altitude they need, they will set their wings and, using the earth's gravity to pull them along, glide to the next opportune lift mechanism — either a thermal or the updraft off another ridge.

Buteo migration is protracted, and it occurs over some part of North America every month of the year. As the Broad-winged Hawks depart their breeding grounds, the vanguard of their fall migration filters past New England and Great Lake hawk-watch points, beginning in mid-August. In the West, Swainson's Hawks, the champions among long-distance buteo migrants, also begin moving south by mid- to late August.

The peak flight for Broad-winged Hawks occurs between September 10 and 20 at northern watch sites (September 17 is the magic date at Hawk Mountain); around September 25, Texas hawk watchers expect their big Broad-winged push

Buteos: soaring juveniles of all species, backlit. Compare size, patterns of translucence, and wing shape. Adults of all species have significantly shorter tails than juveniles, with shorter, broader wings. Red-tailed Hawk **(A)**: well-rounded wings pushed forward at wrist, bulging outer secondaries, dark patagial mark and belly band, light chest, and pale inner primaries. Swainson's Hawk **(B)**: long evenly curved wings with pointed tips; only four slotted primaries; all flight feathers dark. Ferruginous Hawk **(C)**: long, straight-edged, and rather pointed wings; extremely pale; all flight feathers translucent; body strikingly white.

(big flights may exceed 300,000 birds). The peak of the Swainson's Hawk migration typically occurs several days later.

Within a week the birds reach Veracruz, Mexico—a geographic bottleneck formed by mountains to the west and the Gulf of Mexico in the east. Nearly 2 million Broad-wingeds and as many as 1.2 million Swainson's Hawks may pass over Veracruz between mid-September and mid-October—arguably the greatest migratory spectacle in North America, perhaps the world.

In the United States, adult Broad-winged and Swainson's hawks are mostly gone by October, though first-year birds continue to wander south throughout the month. By mid-October, burly cold-weather buteos are advancing on Great Lakes sites, and toward month's end, northeastern hawk watchers fall under the spell of Red-shouldered Hawks—to be followed by the big Red-tailed Hawk push, which may occur at the end of October or during the first week in November (depending upon the fortune of cold fronts, those catalytic conditions that spur migration in the fall).

Among birds of prey, juvenile birds commonly precede adults in the fall, while breeding adults return earlier in the spring. During fall migration, the first day after the passage of a cold front is generally best for viewing large numbers of birds. At some coastal concentration points—most notably peninsulas such as Cape May, New Jersey, and Cape Charles, Virginia, where birds may be bottled up by the con-

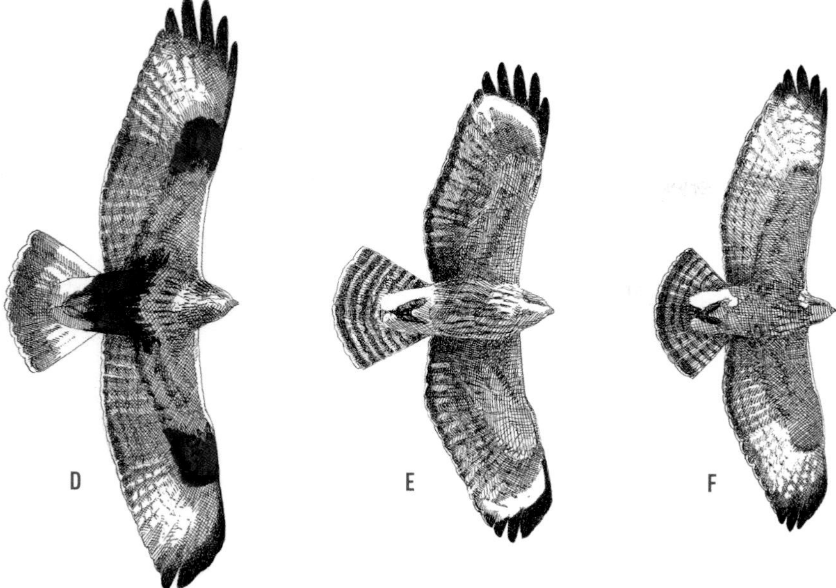

Rough-legged Hawk **(D)**: long, straight-edged, broad-tipped wings; black belly and wrist patches; otherwise rather pale, translucent primaries. Red-shouldered Hawk **(E)**: short wings pushed well forward, with square-cut tips; long tail; translucent slash across wingtip; evenly streaked body. Broad-winged Hawk **(F)**: small, with short, pointed wings; only four slotted primaries; short tail; translucent inner primaries; streaking concentrated on breast.

stricted landmass — the second and third days following the front may have greater numbers.

Fall buteo flights continue in pulses, triggered by each succeeding cold front, growing progressively smaller as the season advances. Heavy snowfall in northern regions sometimes precipitates a late exodus, but by late December and early January it takes a strong cold front to provide even a scattering of weather-pushed, late-moving adult birds — small rewards for a cold day's vigil.

In spring, as early as mid-February, Rough-legged Hawks begin moving north, along with the vanguard of adult Red-tailed Hawks, whose migration will peak with a rush in mid- to late March. Red-shouldered Hawks follow, their migration peaking in early April, just as the mass of Broad-winged and Swainson's hawks rebounds across the Rio Grande. The wave of Broad-wingeds, greatly reduced in number by attrition, fans out and falls out to reclaim territories. They sweep across New England during the third week in April. One-year-old birds follow; their plumages ravaged by molt and wear, these nonbreeding birds, both Broad-winged and, farther west, Swainson's Hawks, continue to trickle north as spring gives way to summer. In mid-July, it is possible to see a few straggling, one-year-old Broad-wingeds and what are presumed to be the first southbound post- (or failed?) breeding adults at the same time in Cape May, New Jersey.

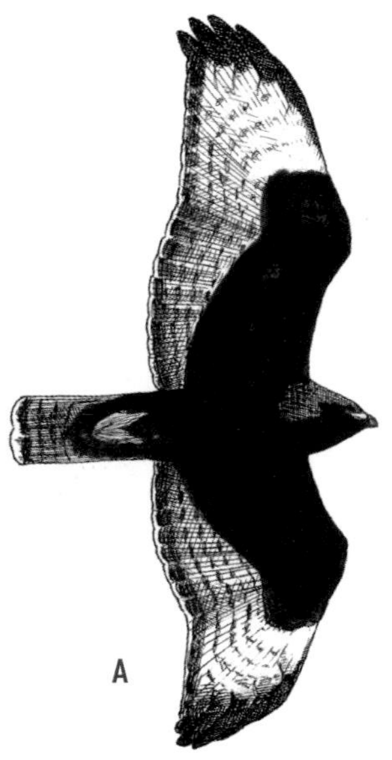

Dark-morph buteo adults gliding. Dark extremes — with blackish body, head, and underwing coverts — are illustrated, but most species are variable and can be rufous or mottled. Juveniles can also be dark and provide more variation. Flight feathers (wings and tail) provide the best clues for ID, as they always appear the same in each species regardless of body and wing covert color. Upperside similar to light morph. Minor variations in plumage may be difficult to see, and identification must often be based on size, shape, and probability. Red-tailed Hawk **(A)**: primaries and secondaries are pale gray with faint banding; tail red (adult) or brown with distinct bars (juvenile). See also Harlan's Red-tailed Hawk. Ferruginous Hawk **(B)**: primaries and secondaries unbarred white, with tiny black tips on primaries (gray on juvenile); white comma at wrist inside black one; tail (mixed white, gray, and reddish) may show faint bars; undertail coverts pale rufous.

IDENTIFICATION

Buteo identification, like the identification of all birds in flight, relies on the integration of a number of hints and clues before a judgment is made. It tends to be more traditional in approach, often relying more upon classic, plumage-based field marks than characteristics linked to structure and flight. There are several reasons for this.

First, buteos are large birds that prefer open areas, and, during migration at least, this is where they are generally seen. Observers viewing them usually have ample time for study and are able to note, even absorb, details associated with plumage.

Second, buteos soar a great deal without flapping. Unless you are already attuned to subtle differences in shape, one soaring buteo looks and behaves pretty much like another.

While the structural differences between species are real, genetics is not the only factor influencing the shape of a soaring bird. Juvenile birds commonly appear rangier or lankier than adult birds. Also, different wind conditions prompt birds to hold their wings in ways that are reminiscent of other species. For example, Red-

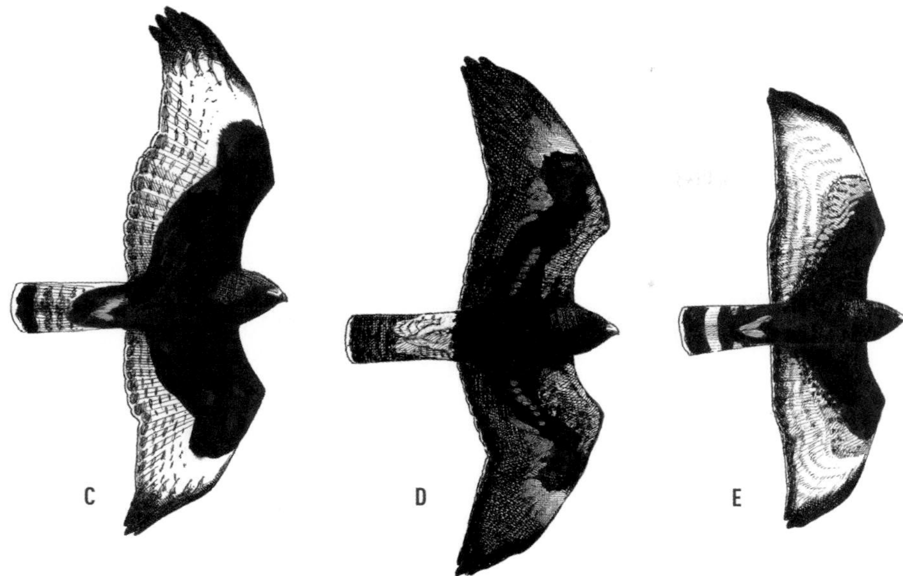

Rough-legged Hawk **(C)**: primaries and secondaries silvery with faint barring; extensive black tips on all; tail barred (adult male) or unbarred pale gray (female and juvenile). Swainson's Hawk **(D)**: dark flight feathers are unique among northern buteos, with pale patch at base of primaries; pale undertail coverts. Broad-winged Hawk **(E)**: all primaries and secondaries whitish with dark tips; tail broadly banded black and white (adult) or brown with several fine bars (juvenile).

tailed Hawks usually soar with a modest dihedral, a slight uplifting of the wings that serves as a useful field mark. But occasionally, when birds are gliding in a strong updraft or when updrafts are scant and the birds are flying full sail with wings fully extended, an approaching Red-tailed Hawk can look flat winged (like a Broad-winged Hawk) or may even show a slightly down-drooped silhouette, a trait more characteristic of a Red-shouldered Hawk.

Third and most important, the field marks associated with buteo plumage and identification are generally reliable and bold enough in most cases to be seen at great distances. If plumage offers a quick and easy route to a correct identification, why take a hard one?

THE GENERIC BUTEO

The typical buteo is a medium to large raptor with a heavy body, broadly proportioned wings, and a comparatively short, broad tail. Wingtips are generally blunt and dark tipped, but among some species that inhabit open country, wings are more tapered and pointed. Upperparts are mostly dark — for most species brownish — and somewhat lightly spangled or mottled. Underparts are usually light with varying amounts of bold, dark streaking, barring, and patches on the underwings and body. *Note: Individuals of several buteo species can be extensively and uniformly*

A juvenile light-morph (Western) Red-tailed Hawk. Over most of North America, and for most of the year, the Red-tailed is the default buteo. JL

dark on the underbody, the underwing linings, the entire underwings, or virtually the entire underparts — these are dark-morph variants. Dark-bodied birds are almost unheard of among other North American raptor groups (Hook-billed Kite, eagles, vultures, and dark Gyrfalcon excepted).

Note: Many earlier books, including the first edition of this one, used the term phase *when referring to the light and dark color variations found in some species. Because this term suggests transition, it has fallen out of favor. Light-morph birds remain light morphs throughout their lives, and dark morphs stay dark.*

In some species, like Rough-legged Hawk, dark morphs are common. In others, like Ferruginous Hawk, they may constitute only a fraction of the population or, in widespread species like Red-tailed Hawk, the distribution of dark-morph birds may vary geographically. Dark-morph Red-taileds, while common across much of the West, are very rare in the East.

Buteos soar frequently and use thermals extensively for lift. They are quick to exploit updrafts off a ridge, particularly on cloudy days when thermals are scarce. Buteos commonly soar and glide more than they flap. For most species, wingbeats are regular, even, somewhat heavy or labored. Commonly given in a series of three to five strokes (sometimes more), wingbeats are usually punctuated by a long glide.

Red-tailed Hawk

The Red-tailed Hawk is a robust large hawk of woodlands, fields, prairies, deserts, and roadsides (even suburbia and large cities) all across North America wherever open ground and elevated perches (most commonly trees) are found. It is the classic buteo, and, over much of its range, for much of the year it is the default buteo.

To simplify the identification process, don't look at a hawk perched on the crossbar of a utility pole, perched on a tree along the highway, or soaring over a woodlot and wonder, "Now, which one of North America's ten buteo species is that?" Ask instead, "Is that a Red-tailed?" Over most of North America, for much of the year, the odds say yes. If it is not a Red-tailed, *then* start considering other possibilities.

In terms of weight, the Red-tailed is among the largest buteos. In the United States, only the Ferruginous Hawk is heavier, while only the Ferrug and the Rough-legged Hawk may exceed the Red-tailed in wingspan and overall length.

The Red-tailed Hawk is the epitome of a raptor, a study in functional diversity. Its hunting habits and habitat preferences are so varied that to single out just one characteristic is to do an injustice to the species.

A Red-tailed can hover like a kestrel or play the forces of wind and gravity against each other and hold motionless in the air like a Short-tailed Hawk—a maneuver known as *kiting*. It can stoop as dramatically as a Golden Eagle and perch-hunt with finesse. Such versatility makes the Red-tailed Hawk as much at home hunting voles in South Amboy, New Jersey, as it is hunting pheasants in South Dakota, or catching snakes in the desert Southwest. And although small rodents are the mainstay of the bird's diet throughout most of its range, the Red-tailed Hawk, like many predators, is highly opportunistic. Any furred, feathered, or scaled creature that is smaller than a groundhog and turns its back on a meal-minded Red-tailed is very likely going to be returned to earth as a pellet.

The Red-tailed Hawk has demonstrated an amazing ability to acclimate itself to the habitat changes fostered by our species (one celebrated pair nests successfully in Manhattan). The bird's high-pitched, two-note, slurred whistle—*kee-yerr!*— is so often dubbed into the soundtracks of movies that it is almost an avian cliché.

Agricultural practices in the Northeast, which cleared bottomland and left hilltops forested, provided ideal Red-tailed habitat. The woodlands harbored the Red-tailed's large stick nests. The wooded edge provided hunting perches, a camouflaged backdrop for a raptor whose underparts replicate a pattern of sunlight and shadow. The fields provided prey.

Dwight David Eisenhower's lasting gift to mobile America, the interstate highway system, further benefited Red-taileds by creating mile upon straight-cut mile of ideal hunting and wintering habitat out of less-Red-tailed-friendly woodland habitat. The bordering woodland edge and navigational aids (i.e., roadway signs) provide perches; wide, grassy highway borders and center divides offer prime habitat for assorted prey; fast-moving traffic and bordering fences ensure that the birds are not disturbed by people.

As propitious for Red-taileds as the leveling of trees in the East was the planting

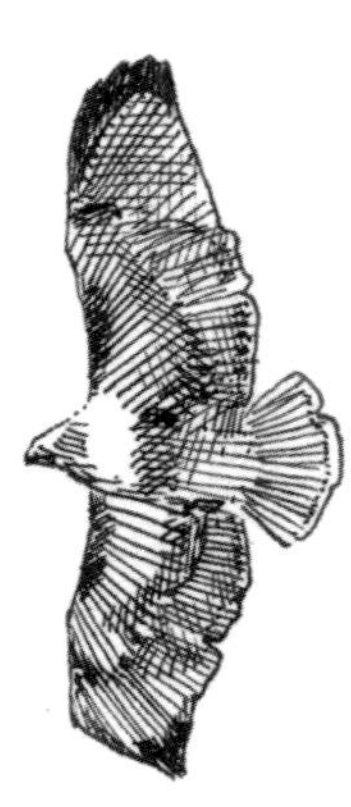

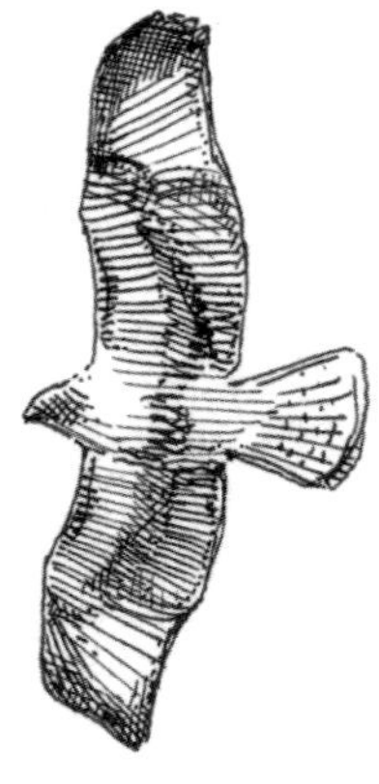

Adult (left) and juvenile (right) Red-tailed Hawk. Adult has shorter, broader wings; shorter tail; slightly blacker markings; often missing or uneven flight feathers; darker (opaque) flight feathers.

of them (or their creosote-soaked remains) in the prairie West. Shelterbelt plantings provided hunting perches and nest sites, while the telegraph, then the telephone, created miles of well-spaced hunting perches where formerly there had been none.

One subspecies or another of Red-tailed Hawk ranges from central Alaska to Hudson Bay, across most of eastern Canada, and south through Mexico and Central America. In winter, northern birds retreat to territories largely south of Canada and the northern plains states. In short, Red-taileds have the largest range of any buteo in the United States. This wide range, along with their healthy population, their propensity to sit in plain view, and their acclimation to our species are what make them the default buteo, and the perfect species on which to begin.

IDENTIFICATION. Study this bird closely. A working familiarity with the Red-tailed provides a reference point for the measure and comparison of all other buteos. This sturdy raptor shows a wide range of size. The largest female Red-tailed Hawk can seem almost as large as a small eagle and weigh twice as much as smaller males. There is considerable variation in proportions as well, generally related to age. Like other buteos, juveniles appear to be rangy, long tailed, and long winged. Adults appear stocky, shorter tailed, and broader winged.

Except for the color of their tails, adult and juvenile light-morph Red-tailed Hawks have fundamentally similar plumage: brown to gray-brown above and light below; an array of pale patches and speckling on the upperwings and back; dark spotting and patterned streaking below.

The basic color on the underparts of juvenile birds may appear white, but most juvenile birds tend toward dirty white or tannish cream. Adults often show a rosy blush across the chest in the East and tawny orange in the West.

Only adult birds have a red tail. The juvenile Red-tailed has a finely barred brown tail that shows white at the base of the tail's upperside. Red feathers replace the brown feathers of juvenile birds when they are a year old (i.e., a second-year bird). Even dark-morph birds and the very pale Krider's Red-tailed (a prairie subspecies) show a reddish tail (pinkish in Krider's). Only the Harlan's form (which breeds in interior Alaska, the Yukon, and northern British Columbia and winters principally in the American prairies) lacks the bird's namesake characteristic. Adult

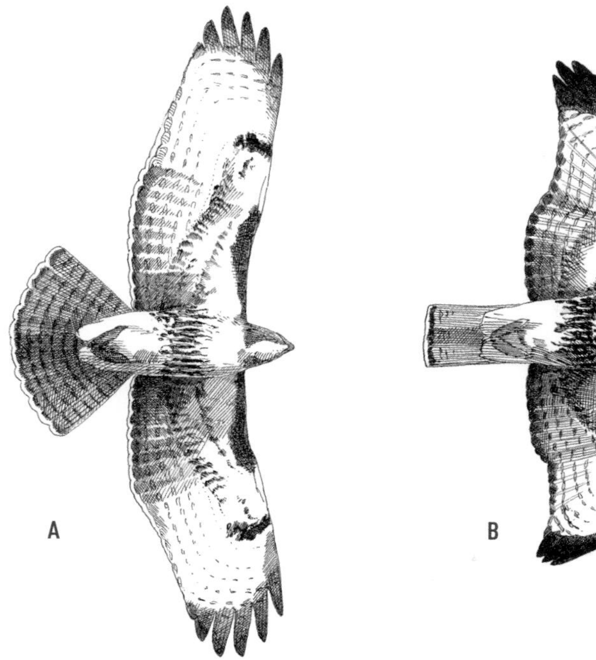

Red-tailed Hawk, underside. Juvenile **(A)**. Adult **(B)**. Large and broadly proportioned; shape variable, but generally wings are rounded and show bulging outer secondaries. Body broad and heavy. Note dark patagium, dark comma, and light area in between that shows as white headlights on approaching bird. Heavily streaked belly band, white chest. Typical individuals of Eastern race (*B. j. borealis*) are shown; variations within this subspecies include birds with heavy black belly band and heavily spotted underwing coverts as well as some with very few dark markings on body, as on Krider's (see p. 25 (A)).

Harlan's Hawks have a gray or white tail, which may or may not show traces of red (see "Subspecies" later in this section).

The rich chestnut red tail of an adult is virtually diagnostic and visible at great distances. The only other buteo species with a true reddish component to the tail is the western adult Ferruginous Hawk, whose tail is mostly grayish or whitish and distally blushed with an orangy red wash.

Caution: When Ospreys, juvenile Red-shouldered Hawks, and Prairie Falcons are backlit by the sun, their tails may take on a pinkish or rosy cast. A male American Kestrel also has a red tail, but this slender robin-sized falcon is unlikely to be confused at any distance with a Red-tailed Hawk.

The long, broad, and blunt-tipped buteo wings appear bulging and muscular, like the overdeveloped arms of a weight lifter. The limbs of other buteos appear straighter, trimmer, and cleaner lined. In a full soar, the hand is often bent forward as if the wing had been broken and healed improperly.

Typical Red-taileds are mostly white below. Adults, western birds in particular, may be blushed with orange, particularly about the chest. Most eastern and western birds boast a dark, patterned band across their underparts — an aptly named *belly band.* It is most prominent in juvenile birds. On some individuals, the belly band

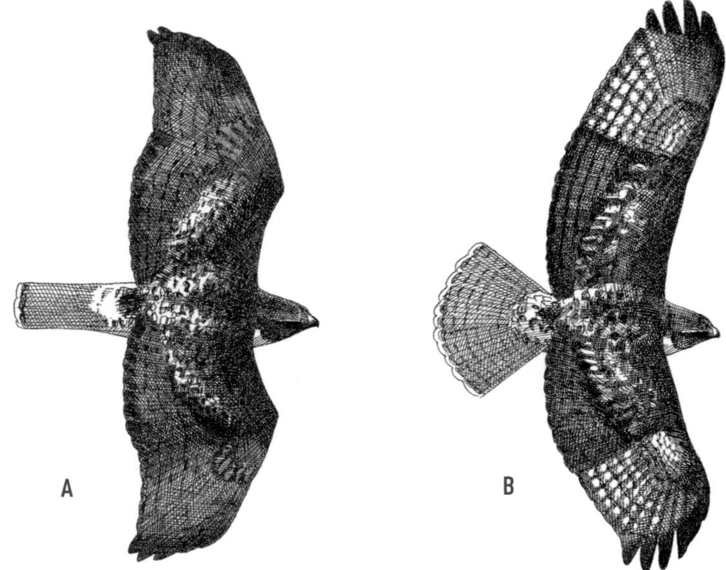

Red-tailed Hawk, upperside. Adult **(A)**. Juvenile **(B)**. Mottled dark brown, with patches of white speckling on greater coverts, scapulars, and uppertail coverts. Adult shows reddish tail with black subterminal band; juvenile has brown tail with many fine bars. Note pale inner primaries on juvenile; lack of pigment here produces patch that looks translucent when seen from below. This patch is less prominent on adults, which are generally darker above.

may be so dark that it resembles the stroke of a black paintbrush drawn across the bird's middle. On other individuals, particularly adults, the belly band may be faint, reduced to spotting or pencil-fine lines, or it may be lacking entirely. In some subspecies (most notably the Fuertes' Red-tailed that is found in the Southwest), the absence of a belly band in adults is the norm.

When present, and even at a great distance, the belly band is arguably the juvenile Red-tailed's most distinctive field mark. Even veteran hawk watchers look for it to make an easy, distant identification.

Caution: Red-taileds are not the only species to show belly bands. Adult female and juvenile Rough-legged Hawks have a very broad black swath across the middle; some juvenile Broad-winged Hawks and Swainson's Hawks have streaking that may coalesce in a band; and immature White-tailed Hawks also have a dark belly. But for the most part, over much of North America, a bird with bright whitish underparts bisected by a ragged dark band across the middle is a Red-tailed Hawk.

The occasional absence of a belly band only accentuates another excellent Red-tailed field mark: the clean whitish breast. Juvenile Red-shouldered, Broad-winged, and Swainson's Hawks and adult Rough-legged Hawks all have streaking on the chest (or at least down the sides of the breast). In the East, a distant buteo flashing a clean white breast is almost certainly a Red-tailed Hawk; in the West, it might be a Ferruginous Hawk; in Texas, a White-tailed Hawk; and in Florida, a Short-tailed Hawk.

Seen head-on, the prominent white chest also serves to accentuate a Red-tailed Hawk's head. With a head more uniformly dark than most raptors, and with dark brown feathering extending down the sides of the face and sometimes onto the throat as well, Red-taileds are sometimes described as wearing a war helmet.

The pale underwings of Red-tailed Hawks are usually more darkly splotched or patterned than those of other buteos. Two distinctive field marks are useful at close range: the bold black comma located at the wrist (carpal area) of the wing, and the dark line running along the leading edge of the arm (the patagium).

The comma is faint or absent on Red-shouldered and Broad-winged hawks but prominent on some Swainson's and Ferruginous, diminishing this field mark's usefulness in the West. In Rough-legged Hawks, the comma is supplanted by a large, dark carpal patch.

Among buteos that migrate, the dark patagium is unique to Red-taileds, present (albeit faintly) on even the pale prairie (Krider's) form. Ferruginous Hawks frequently show rufous-toned spotting or shading along the patagium and, on some individuals, the patagium can be quite dark. However, these dark extremes also tend to be heavily marked along the length of the underwing coverts (a pattern reminiscent of Prairie Falcon). It is precisely the contrast between the dark patagium and paler underwing coverts that distinguishes Red-tailed Hawks and makes this field mark useful at great (though not extreme) distances.

When seen head-on, the white patches between the comma and the patagium on the leading edge of the wings resemble a set of landing lights on an approaching aircraft. This mark is most useful at ridge sites where an eye-level, head-on view of an approaching bird is the norm.

On the wings of juveniles, the primary feathers (particularly the inner primaries) are paler and more translucent than the secondaries. From below, particularly when backlit, they show as a pale square or rectangular window transfusing light through the hand of the wing. Though not unique to Red-taileds, this pale patch is nevertheless prominent — it may involve only the inner primaries and appear as a narrow wedge bisecting the wing, or it may dominate the entire hand out to the dark tips of the primaries.

From above, the pale primaries give the wings of juvenile Red-taileds a two-toned appearance: dark from the base of the wing to the wrist, paler from the wrist to just short of the tip. From a distance, allowing for individual variation, the upper surface of a juvenile Red-tailed's wing appears half-dark, half-light, as though a painter left the job unfinished. The inner half of the wing is finished; the outer portion still needs a second coat.

On the upperwing, the primary coverts are also pale, forming a pale-spotted V stretching from midwing to midwing and across the back of the bird. However, juvenile Red-shouldered Hawks also show a semblance of this pattern.

Western Red-taileds tend to be darker and more heavily marked than most Eastern Red-taileds. Dark-morph and rufous-morph birds, showing dark underparts and wing linings, may be fairly common in the West but are very rare in the East. Rufous morphs usually show the characteristic Red-tailed field marks (belly band, dark patagium, comma) but with less contrast. On dark morphs, these classic field marks are eclipsed (refer to the "Subspecies" section).

Adult rufous- and dark-morph birds still show red tails. On juvenile dark

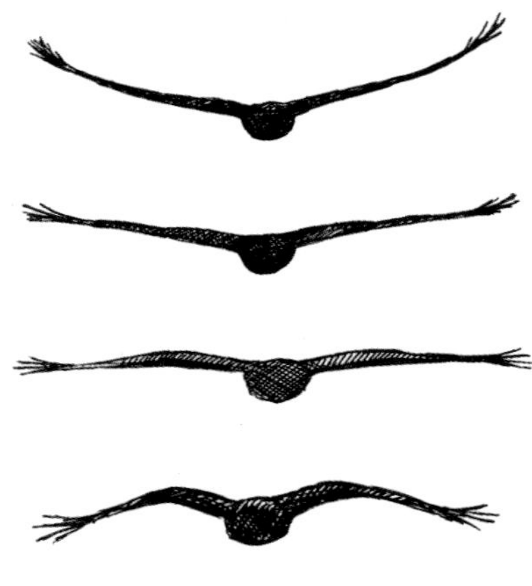

Red-tailed Hawk

Red-tailed Hawk soars with wings in a moderate dihedral, either stiff or upswept; it glides with wings nearly flat or (in a steep glide) pulled in and arched.

morphs, the two-toned upperwing pattern (half of it paler, the other half dark) is still evident.

IN FLIGHT. The Red-tailed Hawk is a master soarer. On days when there is good lift, it may not flap for minutes at a time. When soaring, the bird usually carries its wings in a modest dihedral, less pronounced than that of a Turkey Vulture, Swainson's Hawk, or Northern Harrier. Since the Red-tailed has a heavier wing-loading than these other species, its flight is steadier, not as tippy or rocking.

A soaring and gliding Rough-legged Hawk also has a dihedral, but the wing configuration is more lifted along the arm (as if the bird were shrugging) and flattened at the hands. In contrast, the dihedral of a soaring Red-tailed Hawk is flat along the shoulders and uplifted along the hands. In a shallow glide, the arm rises from the body and flattens along the hand (reminiscent of a soaring Rough-legged Hawk). In a tucked glide (when birds are navigating a strong updraft), the wings are hunched: lifted along the arm, angled sharply down along the hand — a gull-like configuration (but without the gull's curves).

Ferruginous Hawks soar with their wings angled up at the shoulder (like a Rough-legged) but the uplifted wings are straighter, planklike, forming a stiff, modest dihedral. Ferruginous also appears longer winged. Golden Eagles, too, soar with a modest dihedral (that may appear less stiff and more relaxed than in Ferruginous) and likewise appear long winged.

In level flight, the wingbeat of a Red-tailed is measured and heavy, but fluid and usually shallow. The motion appears evenly distributed along the wing — both the arm and the hand work in concert, and nothing is exaggerated. In high winds, if the bird is blown off center, it will right itself by flexing a wing; Red-shouldereds and Broad-wingeds commonly right themselves with a choppy series of wingbeats.

Red-taileds flap less than all buteos except Ferruginous Hawks. They make the most of their weight, moving between thermals at a steeper glide angle than lighter species. When soaring, their bearing is heavier and their turns are slower, more methodical, and wider. In migration, they are late to rise and early to quit the field (except on ridges if the winds remain strong and favorable into the late afternoon).

Red-taileds are capable of hover hunting in place, but over most of North America, only Red-tailed and Ferruginous hawks are adept at kiting (i.e., holding themselves immobile into the wind on set wings like a kite tugging against a string). Rough-legged Hawks can kite, usually for brief intervals, but much prefer to hover. White-tailed Hawks and Short-tailed Hawks are masters at the stop-and-go art of kiting, but are mostly restricted to Texas and Florida, respectively.

Red-tailed Hawk morphs. There is extensive variation. The five views shown here represent the range, but virtually any combination of characters is possible. Adult *B. j. krideri* **(A)**: extremely pale; body and underwing mostly unmarked; head pale; tail mostly white with pinkish wash and dark band near tip; conspicuous translucent patches on inner primaries. Juvenile *B. j. borealis* **(B)**: the typical Eastern Red-tailed; white breast; distinct belly band and patagial mark. Adult *B. j. calurus* **(C)**: the typical Western Red-tailed, usually more heavily marked than *borealis,* with dark throat; rusty streaking on breast and underwing coverts; several dark bands near tail tip. Adult *B. j. calurus,* rufous morph **(D)**: as shown in view **C**, but all light areas replaced by rufous and all markings still apparent. Adult *B. j. calurus,* dark morph **(E)**: body and underwing entirely blackish, no markings apparent. All flight feathers remain as in light morph.

Therefore, over much of North America, any bird that checks its forward momentum and holds its place like a bird pinned to the sky can be called, with near certainty, a Red-tailed Hawk. At any distance it's how a Red-tailed signs its name.

RED-TAILED HAWK SUBSPECIES. With most raptor species, identification stops at the species level. In a few cases, it is possible to identify birds at the subspecies level with a degree of confidence. The Red-tailed Hawk is one of these — or eight of them, if you prefer: *Buteo jamaicensis borealis,* the Eastern Red-tailed; *B. j. abietinus,* the Eastern Canada Red-tailed; *B. j. calurus,* the Western Red-tailed; *B. j. kriderii,* the Krider's Red-tailed (some authorities consider Krider's a color morph rather than a subspecies); *B. j. harlani,* the Harlan's Red-tailed; *B. j. alascensis* of the Pacific coast of Canada and Alaska; *B. j. fuertesi* of the desert Southwest; and *B. j. umbrinus* of Florida. Other subspecies are found south of the United States–Mexico border.

Thus far our discussion has focused largely on the basic or Eastern Red-tailed (*B. j. borealis*), with reference to other subspecies where appropriate or necessary. This section discusses four forms: Western Red-tailed, Harlan's Red-tailed, Krider's Red-tailed, and Southwestern or Fuertes' Red-tailed. All five subspecies have plumage characteristics that distinguish them from the Eastern Red-tailed, and all but Fuertes' migrate substantial distances.

Omitted here are the Pacific Coast Red-tailed and the Florida Red-tailed. These subspecies are more geographically restricted, show less propensity to wander, and share plumage characteristics that are fundamentally similar to the basic Eastern or light-morph Western Red-tailed Hawks.

Observers should bear in mind that many individual Red-taileds (particularly juveniles) cannot be identified to the level of subspecies. Members of each subspecies may interbreed, and intergrades may show an interesting and confusing combination of traits. Furthermore, there is considerable individual variation within each subspecies, so a bird showing some characteristics of a geographically unlikely subspecies is most likely to be just a variant of the common, local population. Bear in mind, too, that partial, even nearly total albinism is not uncommon in Red-tailed Hawks.

A final note to readers: The discussion relating to subspecies is very plumage-centric, and it may be, for some students of raptors, more than they care to assimilate. Don't feel intimidated and don't feel that you must be subspecies conversant in order to identify Red-tailed Hawks in the field. Fundamentally, all these subspecies *are* Red-tailed Hawks; the differences between them are, for the most part, feather-deep. But if you live or travel in the western United States, you will want to be familiar with the array of plumages that will confront you.

WESTERN RED-TAILED HAWK

The typical and most widespread western subspecies of Red-tailed Hawk (and counterpart to *B. j. borealis* in the East), *B. j. calurus* breeds from Alaska and the Yukon south to the Mexican border and the length of the Baja peninsula, and east to the High Plains. As a migrant and wintering bird, it may occur anywhere east of this range.

There are three basic color morphs of *calurus,* which, for reference and simplicity, may be called light morph, rufous morph, and dark morph. Variations seem infinite, but most birds can be assigned to one of these three categories.

Light-morph *calurus* is fundamentally a darker, more heavily printed version of the generic or Eastern Red-tailed. Upperparts are typically darker and warmer toned than on typical Eastern Red-taileds (for you photographers, two f-stops underexposed). Classic Red-tailed field marks are vivid and sharply defined. Underparts (both body and underwing linings), particularly on juveniles and frequently on adults, are more diffusely and heavily marked than on most eastern birds. These markings, too, have a warmer, richer, and redder cast on adults than is common on eastern birds. The face and underparts of adult light-morph *calurus* are often tinged with a buffy orange wash. Those familiar with the less colorful Eastern Red-tailed are invariably struck by how beautiful Western Red-taileds are.

The tails of most adult Western Red-taileds (in all color morphs) are finely barred, or tiger striped. The tail feathers of adult Eastern Red-taileds have no barring (except for a single, narrow dark band near the tip).

The plumage of adult **rufous-morph** *calurus* is a rich, warm, reddish brown — chocolate on cinnamon. Classic Red-tailed markings are visible but less contrasting, particularly on more heavily patterned juvenile birds, which lack the cinnamon-rufous tones of adults. *Caution: Adults of both Swainson's Hawk and Ferruginous Hawk also have an erythrismal, or reddish, morph. Just because it's cinnamon tinged doesn't automatically mean it's a Western Red-tailed.*

On **dark-morph** *calurus,* both adults and juveniles, many of the typical Red-tailed field marks are eclipsed by the overall dark plumage. The tail on adults is classically red. Primaries and secondaries are paler than the coverts and barred on their underside. Adult dark-morph Red-taileds may be distinguished from the very similar Harlan's Hawk by the former's slightly redder overall cast, usually dark (not barred) outer primary tips, and absence of white splotching on the breast (not to mention the red tail!). Juvenile dark-morph Red-taileds are somewhat grainier, less smoothly or uniformly patterned than adults but still less contrastingly streaked and spotted than juvenile Harlan's.

Caution: Adult Red-taileds from the Maritime region of Canada are also typically dark and closely resemble Western Red-taileds. A very darkly and heavily marked light-morph Red-tailed seen in the East is more likely to be of this abietinus *subspecies rather than a Western Red-tailed.*

IN FLIGHT. Little distinguishes the shape of Western and Eastern Red-taileds. Both have broad and relatively long wings that are more contoured and bulging in odd places (outer secondaries, or alulae) than most other soaring raptors.

However, when soaring, Western Red-taileds tend to hold their wings differently. Whereas the wings of Eastern Red-taileds commonly jut upward in a sharp-lined V configuration, Western Red-taileds commonly hold their wings in a shallow, curving, U-shaped dihedral. (This should not be regarded as a distinguishing field mark, but merely an observation.)

HARLAN'S RED-TAILED HAWK

Formerly considered a separate species (*Buteo harlani*), the bird was reclassified as a subspecies of the Red-tailed Hawk in the fifth edition of the American Or-

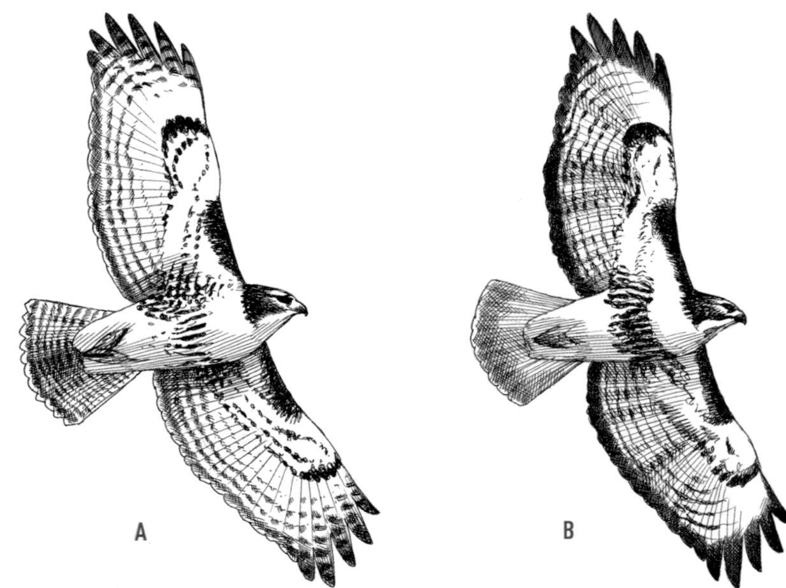

Harlan's Red-tailed Hawk light morph. Juvenile **(A)**. Adult **(B)**. Both differ from other subspecies of Red-tailed Hawk in having very white ground color of body and underwing coverts, with crisp dark markings; white arcs above and below eye.

Harlan's Red-tailed, light-morph adult. Almost uniformly dark above (darker than other Red-tailed subspecies) with light tail; also note white arcs around eye.

nithologists' Union *Check-list of North American Birds,* published in 1973. Subspecies or not, the bird manifests traits that seem to distinguish it from other Red-tailed subspecies.

Harlan's occupies a select and limited range. It breeds in the coniferous forests and bogs of the Alaskan interior, the Yukon, and British Columbia. In winter, the entire population retreats to what was, historically, the tall and midgrass prairies of Kansas, Missouri, Arkansas, Texas, Oklahoma, and southeastern Colorado. A few birds regularly occur as far west as Washington and California, and there is at least one record on the Atlantic Coast.

Make no mistake: this is a long-distance migrant and it could turn up anywhere.

The typical adult Harlan's Hawk is similar to a dark-morph Red-tailed, but the overall color is blacker and cold-

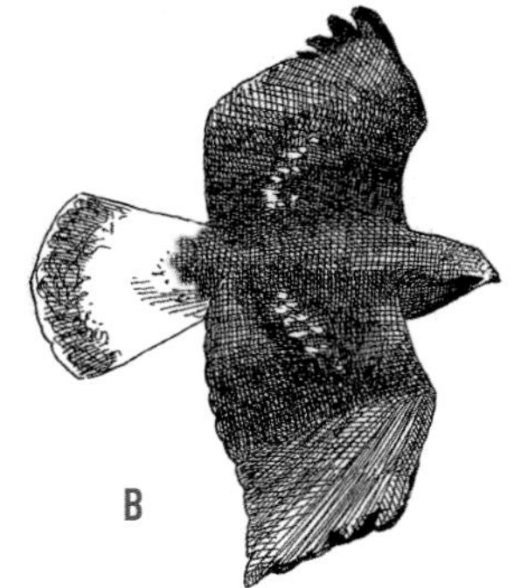

Adult Harlan's Red-tailed Hawk dark morph. Average bird is smaller and slimmer than other Red-taileds, with longer tail and primaries and straighter-edged wings. Body and underwing coverts always dark with white spotting **(A)**, contrasting with silvery flight feathers. Upperside **(B)** dark with a few white spots on coverts. Rump white and tail white, with dusky mottling near tip. The light extreme is illustrated. Other birds have tail entirely mottled. Juveniles not always distinguishable from other dark Red-taileds but always dark with white spotting and checkering on body and underwing coverts; tail banded as on other Red-taileds.

er — the bird has plumage (and even somewhat behavioral and morphological) characteristics that recall a Rough-legged Hawk. The back and underparts are blackish (not brown) with varying amounts of white mottling appearing on the wings, chest, and upper- and underwing coverts. At a distance, the flight feathers of adults are silvery white. At closer range, they are finely barred — *particularly on the tips of the outer flight feathers.* Other dark-morph Red-taileds (and Rough-legged Hawks) usually have solidly black tips on their primaries.

The tails of adults are highly variable. Some are white with dark or rufous streaks running along their length, while at the other extreme are birds whose tails are charcoal with white streaks. The tail patterns of many birds fall somewhere between. Some birds have tails that have narrow black-on-white barring (not streaks). But streaked or barred, light or dark morph, virtually all Harlan's Hawks show broad, dark, subterminal tail bands that range from diffuse to sharply defined.

No matter what the variation, it is a very un-Red-tailed-like tail — the only Red-tailed without a red tail.

The typical juvenile Harlan's is similar to the adult but is more liberally spotted with white on the body, upperwing coverts, underwing coverts, breast, and belly. Some lightly spotted birds appear very dark at a distance, almost black, while heavily spotted birds seem gray. The undertail coverts are distinctly barred, black on white, and appear quite pale on some birds.

The tails of juveniles are always barred. Flight feathers, too, are barred so that the underwings appear a tarnished silver, duskier than adults. The two-toned, pale-arm, light-hand pattern that is characteristic of all juvenile Red-taileds is manifest on Harlan's. Be aware though that most Western Red-taileds also show barred flight feathers.

Harlan's also has a light morph (more common than once presumed) that shows many of the classic plumage points of a Red-tailed (including a dark patagium, commas on the wings, and a crisp, dark belly band) but the overall color is contrastingly blackish and white, without the scattered warm brownish and buffy tones of most other Red-taileds. The adult's tail has the Harlan's signature dark subterminal band, and the outer flight feathers are barred, not dark tipped. Intermediate plumages exist as well.

Adult dark-morph Western Red-taileds are easily distinguished from Harlan's. (The rufous form is, quite simply, a bird of a different color.) Dark-morph *calurus* lacks the pronounced mottling on the breast typical of Harlan's, and the primaries and secondaries are more barred (except at the wingtips), giving the overall wing of the Red-tailed a more tarnished cast at a distance. Note, too, adult Western Red-taileds have *all red tails.* Adult Harlan's do not.

Dark-morph juvenile Western Red-taileds are typically browner overall than juvenile Harlan's Hawks, with more barring on the wings and tail and less white spotting about the body (or none at all). At close range, one can see that the tips of the outer primaries of Harlan's are barred, whereas the outer primaries of dark-morph Red-taileds are typically, but not always, all dark.

Plumages, however, overlap in too many respects to safely distinguish many juvenile birds.

IN FLIGHT. Harlan's Hawks may appear slightly smaller, slimmer, slightly *rangier,* and cleaner lined than a typical Western (or Eastern) Red-tailed. The body is more tubular, the neck longer and snakier. A juvenile Harlan's appears slightly longer tailed.

When gliding into a wind or when riding ridges, Harlan's primaries frequently draw back, making the bird's wings look pointy. Seen head-on, in a glide, the wings show a slight jut at the shoulder that imparts the suggestion of a Rough-legged Hawk.

In fact, taken in sum, the long tail, long neck, and narrower, clean-lined wings (and penchant for active aerial hunting) make Harlan's something of a Red-tailed Hawk with Rough-legged Hawk sympathies.

Harlan's hovers and kites in the wind much like a Red-tailed and, in winter, shows a preference for ridge hunting — hanging low, scanning the flat landscape, with its head constantly swiveling from side to side (even turning to look behind), *as hunting Rough-leggeds do.* Also, it hunts from great heights, turning in tight circles. Harlan's are also up earlier, are active later, and generally seem more energetic than typical Red-taileds (at times, they are behaviorally reminiscent of Short-tailed Hawks). The call of the bird is slightly higher pitched and more monosyllabic than that of the Red-tailed.

KRIDER'S RED-TAILED HAWK

The palest of the Red-tailed group — it can be even paler than the Ferruginous Hawk, another prairie raptor to which it bears more than a superficial likeness — Krider's breeds in the Prairie Provinces of southern Canada and Montana, Wyoming, and the Dakotas. It winters in the southern plains states south to Texas. As a vagrant, it has occurred widely throughout the United States, including Florida, New York, and Cape May, New Jersey, and has also occurred west to the Pacific Coast.

In size and shape, Krider's is a typical Red-tailed (most believe that the bird is simply a color variant of Red-tailed and not a true subspecies); its plumage, nevertheless, sets it apart. From below, the bird is white — adults and juveniles are purged of most of the speckling and spotting that is typical of other light-morph Red-taileds. On adult birds, only the pale semblance of a belly band is apparent (and you have to look closely). Juveniles show a faint spotted outline of this classic field mark. Think of a Red-tailed two f-stops overexposed.

The bird has thin gray, not black, commas and a thin, ghosted patagium. From below, the tail appears white to tan on adults; from above, it appears pale pink. When backlit it remains whitish at the base and pinkish red from the midpoint to the tip. The tails of juveniles are overall pale to white and very faintly barred — more vermiculated, actually.

The dorsal surface of the bird is striking and distinctive — the upperwing coverts and back are mottled white and tan. Except for spare, dark streaking, the head is white. The tail, too, is white at the base, becoming rufous washed or pinkish toward the tip, and has a thin, darker, subterminal band — not at all the striking red tail characteristic of most adult Red-tailed morphs.

Juveniles are like adults, but colder: grayer toned above and lacking the reddish or tawny blush on the finely banded tail. *On juveniles, the large white rectangular upperwing patches that dominate the wing just short of the slotted wingtips are distinctive and obvious.* This patch is reminiscent of the classic two-toned upperwing pattern found on most juvenile Red-tailed Hawks but is much more pronounced. In essence, at any distance, a Krider's is a Red-tailed Hawk with all compass points highlighted in white — white head, white wingtips, and whitish or whitish-based tail (adults with a rosy blush).

SOUTHWESTERN (FUERTES') RED-TAILED HAWK

A handsome, distinctive, but geographically restricted subspecies. It is found primarily in northern Mexico, but its range extends into extreme southeastern Arizona, extreme southern New Mexico, and across southern and western Texas. The most formally and simply attired of the Red-tailed clan, adults are mostly dark brown above (lacking the lavish white mottling typical of other light-morph Red-taileds) and white below, showing little to no streaking — at most only a finely streaked hint of a belly band. Juveniles are like Eastern Red-taileds but lean toward the more lightly marked side of typical.

Fuertes' (and Eastern) Red-taileds typically have a white throat; Western Red-taileds commonly have a dark throat.

1. This adult Eastern Red-tailed Hawk is not just a textbook representative of the species — it is the quintessential buteo across North America, the species against which all other buteos are measured. Note the dark patagium, dark commas at the base of the outer flight feathers, and, in this case, the all-rose-colored tail that appears rich red above. New Jersey, January KK

2. This is a card-carrying adult Red-tailed: a big, compact, richly patterned buteo with an all-red tail. Note, particularly, on this soaring individual, the bulging secondaries (making the wing look muscular) and the short tail. New Jersey, January CS

3. A juvenile Eastern Red-tailed in a slow, shallow glide. Juveniles typically appear rangier than the blocky adults — longer (and more slender) winged and longer tailed. But this individual shows the classic plumage characteristics — dark patagium and commas, dark belly band across the middle, and the squarish, translucent panels showing through. Cape May, October CS

4. A juvenile Eastern Red-tailed, going away. More glide (less soar), therefore pointier wings, narrower tail, and more pronounced curve on the trailing edge of the wing. Cape May, October CS

5. A juvenile Eastern Red-tailed with its wings more tucked as it rides an updraft off a ridge. Classic field marks easily distinguish this bird as a Red-tailed Hawk, but take note of the overall shape: blocky, stocky, robust. New Jersey, November CS

6. A very pale juvenile Red-tailed Hawk encountered in the East. Leucistic, partial, and full albino birds turn up with some regularity in this species. This individual, seen in Cape May in November, had Krider's-like characteristics below, but not above. Point here? Expect a vast array of plumage variation among Red-tailed Hawks anywhere in North America, particularly in the West. CS

7. An adult light-morph Western Red-tailed Hawk from above. Much like Eastern Red-tailed, but with slightly darker upperwings and body. Atypically, the tail appears to be unbanded in this individual. Nevada, September JL

8. A juvenile light-morph Western Red-tailed Hawk from above, showing the pale outer wing panels and mottled upperwing coverts typical of juvenile Red-tailed. Note the finely banded brown tail. Some birds are lighter, some darker than this. The "wings jutting forward at the wrist" posture is classic for soaring Red-tailed. Utah, October JL

9. A classic light-morph adult Western Red-tailed soaring, showing expected fine bands in the red tail (fine tail bands are absent in Eastern Red-tailed). Arizona, September NH

10. A gliding juvenile light-morph Western Red-tailed. Young Western Red-taileds tend to be more heavily marked than Eastern, but show all the classic Red-tailed features (including the pale wing panels bleeding through this backlit bird). Utah, October JL

11. RIGHT: A light-morph adult Western Red-tailed, but a richly rufous one. Variation is considerable. Utah, October JL

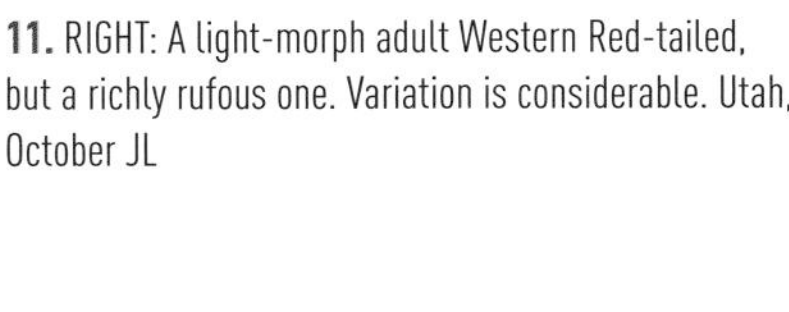

12. BELOW: A soaring rufous-morph adult Western Red-tailed Hawk. Note the darker patagium and belly band. Shape is spot-on for Red-tailed. Utah, October JL

13. A true (full) dark-morph adult Western Red-tailed, gliding as it rides a mountain ridge updraft. Yes, there are other dark-morph buteos, but not with a red tail (or with such a classic Red-tailed shape). Utah, June JL

14. A juvenile Western Red-tailed dark morph (the pale extreme), banking in a full soar. Note the variation in plumage when compared to #15. Although this individual has barred primaries, it is not a Harlan's Hawk, as it fledged from a Utah nest. Utah, October JL

15. LEFT: A juvenile Western Red-tailed dark morph (the dark extreme) in a shallow glide. Note the shape and pale wing panels. Nevada, September JL

16. BELOW: An adult Fuertes' Red-tailed Hawk in a soar. This Southwesterner is virtually identical to a very pale Eastern Red-tailed. Note the near absence of a belly band and the unbanded red tail. Western Red-taileds (and particularly Fuertes') seem to show more pronounced or U-shaped dihedral (with flat arms and curved, uplifted wingtips) than do Eastern Red-taileds. Arizona, February NH

17. A classic adult dark-morph Harlan's, soaring. Dark morph is more typical of Harlan's (although it is possible that many light-morph birds go unrecognized). Note the overall clean-lined, slightly lankier shape of this bird. Overall tones are cold. White mottling on the blackish breast is typical. Tail is highly variable, but birds with smudgy bands on grayish tails are common. Alaska, April JL

18. An adult dark-morph Harlan's in a glide. Note the absence of a broad dark band on the tip of the gray tail (therefore not a Rough-legged Hawk). While many (most?) Harlan's show banding on the tips of the outer primaries, some adults do not. Alaska, April JL

19. An adult dark-morph Harlan's (but a somewhat mottled individual) in a soar. Note the touches of red in the otherwise gray tail — also a trait that is very common in this subspecies. Alaska, April JL

20. A juvenile dark-morph Harlan's in a full soar. Note the pale mottling on the breast, banded primaries, and darkish subterminal tail band — not as broad as on a Rough-legged Hawk. But the overall shape, particularly the clean lines of the wing and planklike symmetry, does hark to Rough-legged. Notice that the pale wing panels common in juvenile Red-taileds are present, but not prominent in this image. Utah, October JL

21. A light-morph adult Harlan's Hawk in direct flight. This beautiful bird has a lot of classic Red-tailed features (except for the curious gray patterning on the flight feathers). But note the pale gray tail! That's a classic Harlan's feature. Alaska, April JL

22. The dorsal view of a Harlan's, banking. It truly is a Red-tailed of a different color. There is immense variation among individuals, but here the dark-streaked white tail (with slight touches of red at the base) is fairly typical. Utah, March JL

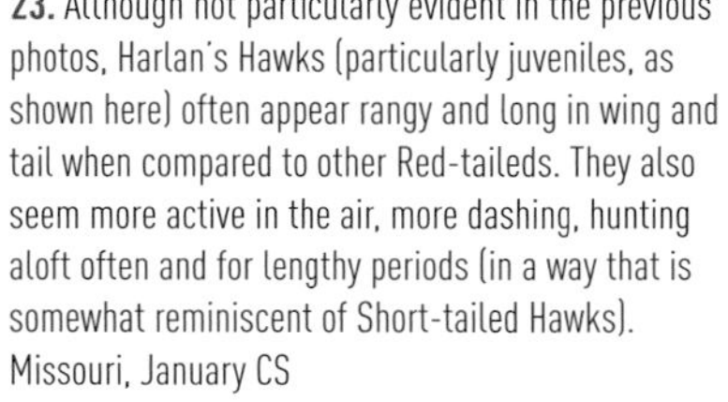

23. Although not particularly evident in the previous photos, Harlan's Hawks (particularly juveniles, as shown here) often appear rangy and long in wing and tail when compared to other Red-taileds. They also seem more active in the air, more dashing, hunting aloft often and for lengthy periods (in a way that is somewhat reminiscent of Short-tailed Hawks). Missouri, January CS

24. An adult Krider's Red-tailed Hawk. Classic Red-tailed shape but almost wholly white below with a white head and slightly blushed white tail. Note, however, typical dark Red-tailed Hawk patagium and dark comma at the base of outer flight feathers. North Dakota, July JL

25. An adult Krider's Red-tailed Hawk, hovering (as all Red-taileds easily do, although kiting or stalling is more the Red-tailed's fashion — particularly in the windy open spaces and strong thermals of the American West). An even paler individual than #24. North Dakota, July JL

26. An adult Krider's Red-tailed from above, showing fairly dark upperparts but a very pale tail. Some adult tails are even paler above. North Dakota, July JL

27. A juvenile Krider's Red-tailed, dorsal view, showing the classic pattern of white head, white wing patches, and whitish tail. Also apparent, in fact accentuated, are the pale braces along the upperwing coverts found on most Red-taileds. Oklahoma, January JLISH

28. An apparent juvenile Krider's Red-tailed showing a pale head, but the pale wing panels have disappeared because of molt. And . . . is that a standard-issue Red-tailed Hawk tail feather molting in? The lines are often blurred between Red-tailed Hawks subspecies. North Dakota, June MOB

29. Krider's Red-taileds sometimes wander east. Here is an apparent full Krider's (adult) over the Florida Prairies. Notice the pale face and tail, and the very muted pattern below. Florida, February KK

Red-shouldered Hawk

The Red-shouldered Hawk is a forest buteo — the buteo that *thinks* it's an accipiter. It is at home in the forested hillsides of New England, woodlots in Ohio, the cypress swamps of the South, the piney woods of Texas, riparian corridors in Kansas, and oak-mantled residential areas in California. It is absent across the tree-deprived prairie states, most of the American West, and all but southeastern Canada. Five subspecies are recognized but three of them, *Buteo lineatus lineatus, B. l. texanus,* and *B. l. alleni,* are essentially indistinguishable from one another and occupy most of the eastern states. Southern Florida birds, *B. l. extimus,* are paler than the other eastern subspecies. The geographically isolated California Red-shouldered, *B. l. elegans,* is more richly colored, boasting an unbarred rufous bib and broader white bands on the tail. California juveniles also differ greatly from their eastern relatives by being, essentially, rough-cut versions of the adults. In other subspecies, juveniles are brownish above and whitish below with breasts more streaked than barred.

Regional plumage variations notwithstanding, in the hawk-watching arena these shades of difference diminish to insignificance. Wherever it occurs, the Red-shouldered Hawk can be treated as a single entity because all of its idiosyncratic traits hold true.

Red-shouldereds are perch-hunting raptors. Targeting mostly small rodents, during warmer months they augment their diet with insects, reptiles, amphibians, and crawfish (where they occur). During the winter, small birds figure more prominently in the Red-shouldered's diet (and in some places, the birds haunt backyard feeding stations). The talons of the Red-shouldered are significantly smaller than those of the Red-tailed Hawk. Prey larger than a gray squirrel is the exception, not the norm.

As forest birds, Red-shouldered Hawks do not favor the open areas frequented by many other species of buteos, so they are less easily seen. In winter, the birds frequently perch on the sunny edge of fields, most commonly selecting perches at middle heights, not high in the tree like Red-tailed Hawks commonly choose. In California, the birds are frequently seen along riparian (and traffic) corridors. In Florida, Red-shouldereds are common, well acclimated to people, and found in a variety of wooded and open habitats, including suburbia.

When on their breeding territories, the birds can be very vocal, alerting observers to their presence with a loud, ringing *kee-ah! kee-ah! kee-ah!* Juveniles and nonbreeding adults are mostly silent, particularly during migration.

IDENTIFICATION. The Red-shouldered Hawk is a medium-sized buteo about three-quarters the size of a Red-tailed Hawk. The body is slighter than a Red-tailed's and is slimmer and tube-shaped. Overall it appears trimmer, cleaner lined, and more clean-cut. If the ubiquitous Red-tailed Hawk can be likened to "jes' plain folk," the Red-shouldered Hawk is gentry.

Red-shouldereds are beautiful; adults, in particular, are striking. Perched, with wings folded, the bird appears to be wearing a dark cloak embedded with an array of white spangles. The bird's head, underparts, and especially its namesake

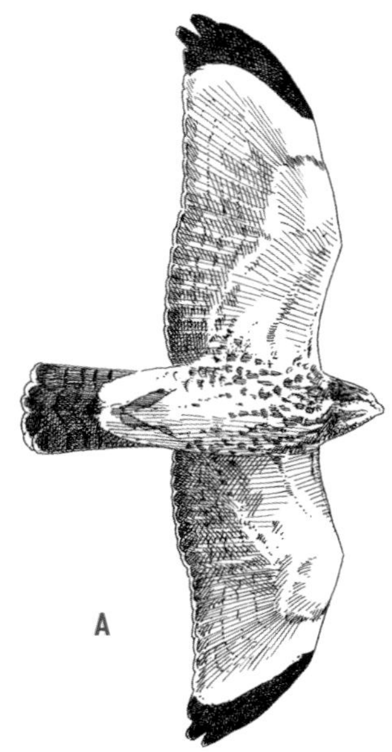

Red-shouldered Hawk, underside. Juvenile **(A)**. Adult **(B)**. Medium-sized, rather stocky; wings clean edged, rounded. In all plumages the translucent slash just inside the black primary tips is prominent. Bird is otherwise distinguished from Broad-winged by larger size and broader wingtips; juvenile also by (usually) uniformly streaked breast; adult also by narrow white bands on tail, barred primaries and secondaries, rusty barring on wing coverts.

shoulders are rufous-splashed. The tail is crisply contrasting — narrow white bands set on black.

The smaller adult Broad-winged Hawk is not so smartly patterned nor so ruddy complexioned, and it lacks the Red-shouldered's rust-colored epaulets. The tails of the two species are superficially similar, but the white bands on Red-shouldereds are narrow — chalk lines drawn on a black slate. When an adult Broad-winged is sitting, its single visible white tail band is broad and distinct.

Red-shouldered Hawk, juvenile California subspecies. Much more adultlike than immature of eastern subspecies. Range does not overlap with eastern birds. Shows narrow white bands on dark tail, whitish crescent window on wingtips, and barred flight feathers like adult. Underparts heavily marked with streaks and bars, underwing coverts dark.

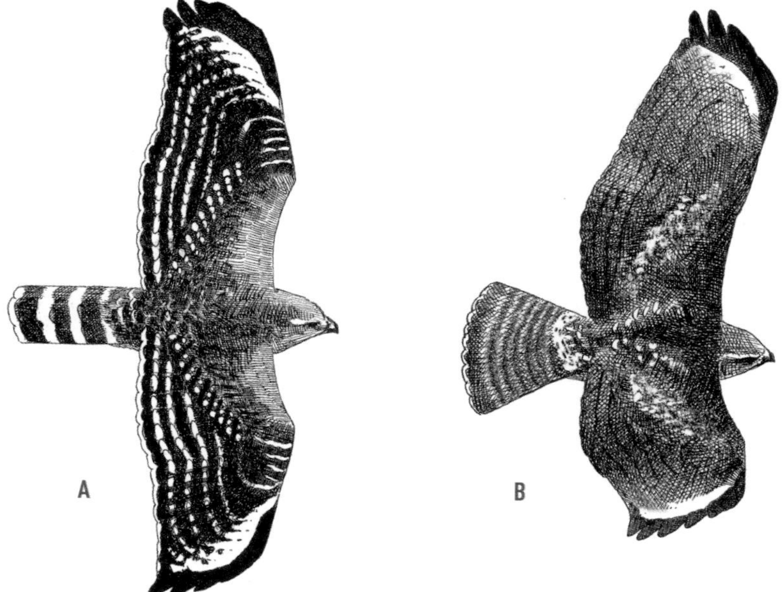

Red-shouldered Hawk, upperside. Adult **(A)**. Juvenile **(B)**. Adult strikingly patterned, with rufous shoulders and black-and-white barring. White crescent on wingtip produces translucent window below. Juvenile brown with some white spotting on coverts, scapulars, and uppertail coverts; bold tawny crescent across wingtip distinguishes this species from all others.

Except in California, juvenile Red-shouldereds lack the distinctive plumage of adults. Young Red-shouldereds are brown above (variously spangled with warm buff and white spots) and cream-colored below with streaking (or barring in some subspecies) that is *particularly heavy on the chest.*

IN FLIGHT. Wings are long, symmetrical, and clean edged — lacking the muscular bulges seen on Red-tailed Hawk. The leading edge is straight. The trailing edge curves gently (most noticeably on adults) or hardly at all (usually characteristic of juvenile birds). The wingtip is unusually blunt, cut straight on an angle. Seen from below, the wing of a Red-shouldered Hawk suggests a rectangular plank — a long plank on a juvenile, a shorter plank on an adult. In a full soar, the entire wing sweeps forward as if the bird were reaching out, arms wide, to embrace the sky.

A juvenile's tail is long for a buteo but still buteo-broad. The adult has a rather short tail, and it appears even shorter when a bird is soaring and the tail is fully fanned. The tail is blunt, not notched, unless a bird is molting.

Putting the parts together, you are left with two birds (more correctly age classes) that have markedly different shapes. Long-winged, long-tailed juveniles appear lanky and rangy, recalling an accipiter. Shorter-winged and shorter-tailed adults appear stocky and compact. In side-by-side comparisons, soaring birds might easily be regarded as different species.

But while adults and juveniles are differently proportioned, both are *nicely*

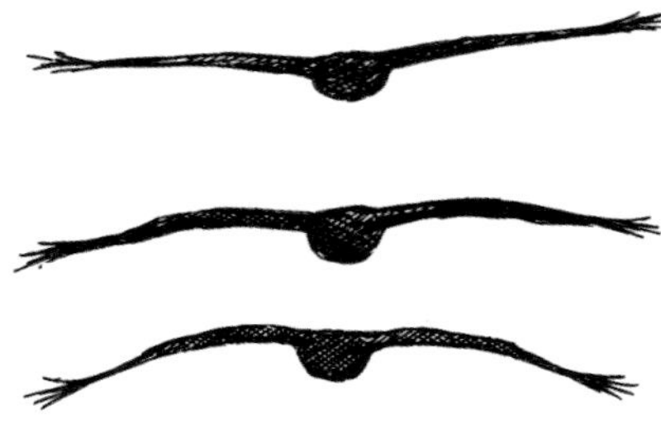

Red-shouldered Hawk

Red-shouldered Hawk soars with wings flat or slightly raised or drooped and glides with wings arched or cupped.

proportioned, with trim-cut, clean-lined, symmetrical wings that are more planklike (juveniles) or paddlelike (adults) than tapered. Their silhouettes are different, but both are distinctive — very unlike other buteo species.

The Red-shouldered Hawk's classic (and arguably best) field mark is the *crescent-shaped, translucent stained-glass-like window* that contours the tips of the bird's wings. The absence of pigmentation near the base of the outer flight feathers allows light to pass through the wing, illuminating the window. The resulting narrow, pale crescent is visible from above and below. It is tawny on juveniles in the East, white on eastern adults and on adults and juveniles alike in the West, and visible at great distances. *Caution: Other buteos commonly show translucent ovals or rectangular patches on the wings. They are not crescent-shaped and are usually relegated to the inner primaries.*

Another caution: Molting Red-tailed Hawks, Broad-winged Hawks, and other species may occasionally display tawny crescent-shaped windows that recall Red-shouldered Hawks. The molt of one or more middle primaries allows light to transfuse through the gap. During fall migration, most species are not molting flight feathers. In late spring, late-migrating subadult or second-year Broad-winged Hawks frequently have missing primaries and often show what appears to be a crescent

Broad-winged Hawk (left) and Red-shouldered Hawk (right). Juveniles in spring at about one year old undergoing molt to first adult plumage. As Broad-winged loses the middle primaries it can develop a pale crescent across the wingtip very similar to unmolted Red-shouldered, as worn translucent outer primaries contrast with new, more opaque ones. Note size, tail pattern, and dark trailing edge on wing, and use extra caution when identifying molting birds in May and June.

window. It is not. It is a gap in the wing caused by a missing feather, which can be confirmed by noting the broken symmetry on the trailing edge of the wing.

Juvenile Red-shouldereds are variously streaked (or barred) on the chest. Young, light-morph Red-taileds are streaked across the middle but comparatively clean chested. Young Broad-winged (and many young Swainson's) Hawks are most heavily streaked down the sides of the face, throat, and sides but are often clean chested.

Juvenile Eastern Red-shouldereds are brownish gray above (other subspecies are grayer). At a distance, the infusion of buffy and ruddy feathers in the back, tail, and wings gives the bird a warm brown cast. Juvenile and adult Eastern Red-taileds project a cold brown cast at great distances.

The Red-shouldered is animated, high-strung, and considerably more active in flight than a Red-tailed Hawk — the bird that it is most commonly confused with in migration. In direct flight, the wingbeat is quicker and lacks the fluid quality of a Red-tailed's. The beat appears stiff, as if the bird were batting the air. The flight looks as if a buteo were trying to imitate the more sputtery flight of an accipiter. *Rule of thumb: If you initially identify a distant bird as a Northern Goshawk, and if, after closer study, it turns out to be a buteo, you are probably looking at a Red-shouldered.*

Juvenile Red-shouldered Hawk in full soar. Wings thrown forward, with pale wingtip crescent obvious.

Because the Red-shouldered Hawk weighs less than a Red-tailed, it is less able to buck high winds. When it is thrown off balance, it will right itself by flapping frantically. A Red-tailed will simply flex a wing or make one or two methodical, corrective flaps. In migration, on ridges, and particularly on days characterized by strong winds and updrafts, the lighter Red-shouldered Hawk will commonly ride higher and on the off-wind (leeward) side of the ridge (avoiding the turbulent, buffeting air that a larger, heavier bird — like Red-tailed or eagle — can accommodate).

Juvenile Red-shouldered (right) and Northern Goshawk (left). Shape and size are very similar in these species, but the pale wingtip crescent of Red-shouldered and narrow hands of goshawk are useful clues.

Red-shouldereds typically soar on flat wings and glide with wings that have a down-curve or bow, as if the bird were cupping the air. Sometimes the wings of birds seen head-on seem positively *bowed* — raised at the shoulder and curved along the wing's length.

1. A soaring adult Red-shouldered Hawk, showing the classic short, slightly paddle-shaped wings. The very rich orange color is typical of California birds. Overall, a fairly clean-limbed, compact bird of prey. California, February KK

2. An adult Red-shouldered Hawk of the pale Florida subspecies. Note the broader white bands in the tail than #3, and how much paler the bird is overall when compared to #1 or #3. Red-shouldereds have variable plumage, but a classic profile: soaring birds thrust their wings forward as if reaching out to embrace the sky. Florida, January KK

3. An adult "eastern" Red-shouldered with narrow white bands in the tail. This rust-colored bird shows the adult plumage seen most widely over the Red-shouldered's range, and this plumage is often enjoyed at eastern hawk watches. This bird shows nicely the blunt-tipped wings typical of Red-shouldered and the reaching or "hugging" shape as the wings are thrust forward. Cape May, November CS

4. A juvenile Red-shouldered Hawk gliding. Note the translucent "windows" at the tips of the wings. The streaking covers the entire breast (not concentrated along the throat and sides of the breast like Broad-winged, or across the belly like light-morph Red-tailed). Note, too, the absence of a dark patagium and dark comma at the wrist. Michigan, April JL

5. A soaring adult Red-shouldered from above, showing its namesake characteristic. The white crescent-shaped "windows" are apparent just short of the wingtips. Eastern Red-shouldereds are similar, but the reddish tones are paler, more rusty (also see p. 10). California, February KK

6. A juvenile Red-shouldered Hawk in a glide. Note that juveniles commonly appear lankier and less compact than adults (and particularly longer tailed). This bird is showing the crescent windows and spotted breast. Cape May, November CS

7. A juvenile Red-shouldered in a full soar — reaching forward with its wings. Crescent windows are not prominent in this bird at this angle. Cape May, November CS

8. A juvenile Red-shouldered Hawk "hugging the sky." Crescent windows, visible above and below, are found on adults and juveniles, but are often most prominent on juveniles. Crescents are most prominent when birds are backlit, and they can be visible at some distance. Cape May, November CS

9. A juvenile Red-shouldered Hawk gliding. Gliding birds typically droop or cup their wings. Think "bowed wings," with arm raised and hand dropped. (Red-taileds glide with a slight dihedral; Broad-wingeds glide with wings angled slightly down.) Note, again, pale crescent windows — a field mark that rarely disappoints. Cape May, November CS

Broad-winged Hawk

If a seasoned hawk watcher in Massachusetts were asked to name the best time to watch hawks, the answer would likely be: "Broad-wingeds peak about the fifteenth of September." In many places in the East (if "East" can be stretched to encompass coastal Texas), hawk watching is almost synonymous with Broad-winged watching. And why not? The annual fall exodus and spring return of the Broad-winged Hawk ranks among the great spectacles of nature. Even veterans who have witnessed the pageant for many seasons look forward to the next and pray that *this time,* this time, fortune will place them in the path of the big flight, on the big day.

This quantitative threshold can mean 10,000 birds in a day, as it does across much of New England; 50,000 birds in a day at some Great Lakes hawk-watch sites; or even 300,000 to 500,000 birds in a single day, as has happened in Texas.

Swainson's Hawk, too, travels great distances in vast numbers. But with a migration route largely free of concentrating leading lines until the birds reach Texas and Mexico, the birds have yet to build the ardent following that Broad-winged Hawks enjoy.

The Broad-winged is a small buteo of eastern deciduous and mixed deciduous-coniferous forests. Breeding primarily east of the prairies, the species ranges west along the forested northern edge of the Canadian prairies to, perhaps, British Columbia. Across most of the American West, it is a rare or casual visitor. In winter, except for a small wintering population in Florida, the entire population withdraws to Central America and northern South America.

Broad-wingeds appear docile by raptorial standards. Even the bird's expression seems gentle. Perch hunters of forest interiors, Broad-wingeds capture and consume small mammals and a few birds. But cold-blooded prey such as insects, frogs, snakes, and toads figure prominently in their summer diet and, perhaps for this reason, Broad-wingeds leave early in the fall and return late in the spring — earlier and later than any other eastern raptor except Mississippi Kite (another bird whose diet is insect-heavy). Unlike the Red-tailed Hawk and Red-shouldered Hawk, which merely vacate northern portions of their ranges and withdraw deeper into southern reaches, in winter the entire Broad-winged population vacates its North American breeding range.

In migration, the bird is highly gregarious, the quality that has earned this species its popular acclaim among the hawk-watching community. Migrating aggregations range from several individuals to many thousands of birds. Migrating birds stream across the landscape, seeking out thermals. The birds cluster in the column, soaring in circles, riding an invisible elevator aloft. These milling aggregations are called *kettles* in many places, *boils* in others — apt descriptions insofar as the milling birds recall water roiling in a boiling kettle.

Small and pale below, individual Broad-wingeds and streaming groups (i.e., well-spaced birds flying between thermals) can disappear when set against a cloudless sky. But when hundreds or thousands of birds cluster in a kettle, the swirling concentrations have enough mass to project themselves at distances exceeding four miles. The echo of these kettles can even be seen and monitored on weather radar.

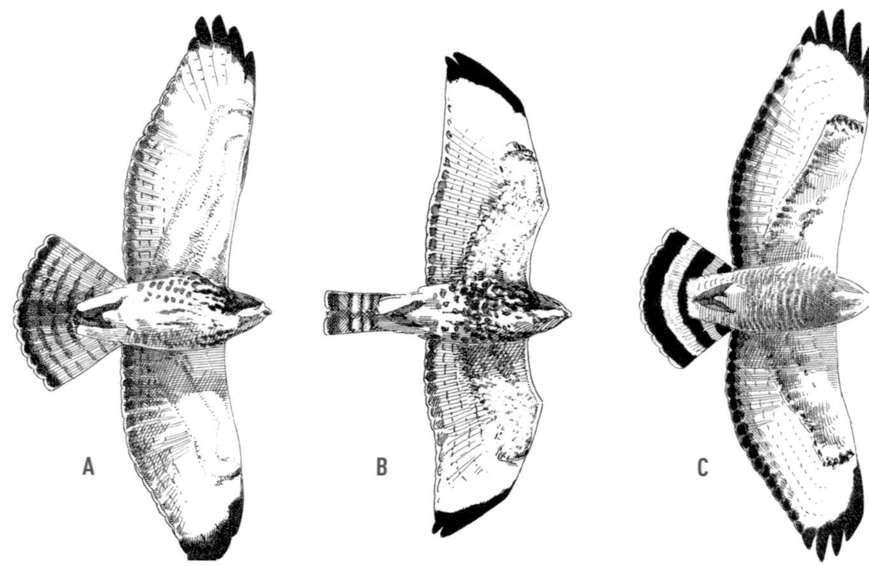

Broad-winged Hawk, underside. Typical juvenile **(A)**. Heavily marked juvenile **(B)**. Adult **(C)**. Small, chunky; wings broad, clean edged, pointed. Tail broad when fanned in soar but narrow and square tipped in glide. Juveniles differentiated from Red-shouldereds by shape, lack of pale window, and streaking and spotting concentrated on side breast and flanks (so that breast remains relatively clean), sometimes forming belly band as on **(B)**. Underwing typically unmarked white **(A)** to lightly spotted **(B)**. Tail bands variable but may be helpful in individuals with conspicuous dark subterminal band as in **(A)**; Red-shouldereds normally have uniform bands. Adults have faint whitish markings and a conspicuous black border on underwing and show one broad white band across tail. For the rare dark morph, see p. 17 **(E)**.

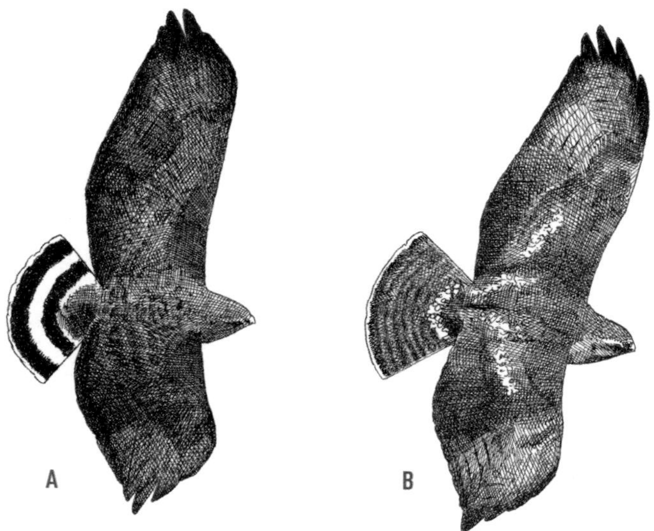

Broad-winged Hawk, upperside. Adult **(A)**. Juvenile **(B)**. Adult uniformly brownish with banded tail. Juvenile very similar to Red-shouldered Hawk but lacks tawny crescent; may show pale inner primaries, light brown tail.

IDENTIFICATION. A Broad-winged Hawk is a small, stubby buteo that is barely larger than an American Crow. A high-soaring adult is easily recognized by the single broad white band bisecting a black tail. At very close range or when birds are harshly backlit, a second, narrower and usually broken band may be noted. By comparison, the tail of an adult Red-shouldered shows multiple narrow white bands. The tail of juvenile Broad-wingeds is brown above, buff below, and banded by a number of narrow dark bands. The subterminal band is conspicuously wider, darker, and more prominent than the other bands. At close range, this cue helps distinguish a juvenile Broad-winged from a juvenile Red-shouldered Hawk, whose subterminal band is marginally wider than other tail bands but does not stand out or seem noticeably different.

Juvenile birds are brown above (with some spare, pale mottling), creamy below, and brown streaked on the body — the same basic plumage pattern found on many young raptors. On a juvenile Broad-winged, however, the streaking is heaviest on the sides of the throat, the sides of the breast, and across the belly. Unlike on most young raptors, the breast on most Broad-wingeds is less heavily marked and may in some cases be spotless. On some individuals, the pattern of streaking gives young Broad-wingeds the semblance of a belly band — a field mark most characteristic of Red-tailed Hawks. Most Red-taileds, however, are symmetrically streaked — a distinct, broad swath that cleanly bisects the underparts. The pseudo–belly band of some juvenile Broad-winged Hawks is an overwash of the heavy streaking that extends down the throat and sides of the bird and meets across the belly. It is usually sparse and not well defined — just a lot of dark streaking that somehow missed the center of the chest.

Dark-morph Broad-wingeds are very rare east of Texas (perhaps 1 in 50,000 birds). They are more common in the western limits of the species' range. At the Golden Gate Hawk Watch near San Francisco, a dark-morph Broad-winged will appear for every 30 to 50 birds seen.

Adult dark-morph Broad-wingeds have the same distinct tail pattern as the standard Broad-winged and silvery white flight feathers on the underwings. Both dark adults and juveniles can easily be distinguished from most other dark buteos by their compact proportions and distinctive shape.

IN FLIGHT. In a full soar, a Broad-winged's wings are short, buteo-broad (but not overly so), and appear tapered along the leading and trailing edge to a point. They jut out at a right angle, perfectly perpendicular to the body, and the lines are superbly clean, devoid of bulges or bumps. In a soar, the wing resembles a low-burning candle flame. In a glide or when soaring in a partial tuck, the leading edge of the wing curves back and the trailing edge straightens. The wing then resembles a paring knife.

The underwings of both juvenile and (particularly) adult Broad-wingeds are very clean, white, and less marked than those of most buteos. The whiteness is accentuated by the broad, dark outer border that defines the outer primaries and darkens the trailing edge of the wing. The impression is that of stretched white canvas bordered by a dark frame. It is very pronounced on adults and less pronounced but apparent on juveniles.

During a soar, the tail opens very wide so that both it and the already stubby

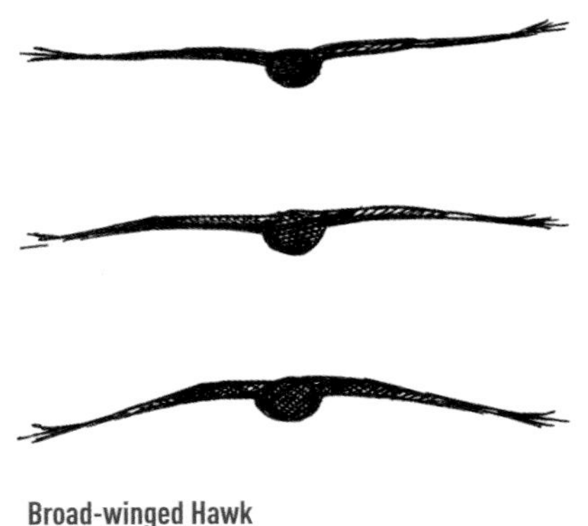

Broad-winged Hawk

Broad-winged Hawk soars with wings flat or slightly raised and glides with wings very slightly bowed or angled down. The wingtips are more pointed, with less obvious fingers, than on other buteos.

wings appear shorter still. When the tail is closed, it is surprisingly narrow and long and can appear deceptively unbuteolike — in fact, it is somewhat reminiscent of Sharp-shinned Hawk, an accipiter. Like Sharp-shinned's, Broad-winged's tail is frequently notched at the tip and may flare slightly.

The wings of soaring birds are held horizontally and flat, not uplifted or drooped. In a glide, between thermals (when wings are drawn in or tucked) or when the bird is riding a strong updraft on a ridge, the wings are angled stiffly downward.

Also, when an adult is seen head-on, as is often the case when ridge watching, the bright yellow cere and white feathers around the base of the bill may be visible at a considerable distance.

As mentioned earlier, Broad-wingeds are highly gregarious during migration, traveling in flocks. In much of the East, spotting a large flock of buteos automatically means Broad-winged Hawks, especially during the species' very narrow migratory window. (Other birds also travel in large thermal-exploiting flocks: Swainson's Hawks in parts of the West, and Mississippi Kites in the Southeast, as well as Sandhill Cranes over much of the continent, Anhingas in the Southeast, and American White Pelicans in the interior and Florida.)

Broad-winged Hawks are more thermal dependent than most other species. Flights do not commonly commence until morning thermal activity begins, and they shut down as soon as thermal production ceases in late afternoon. When gliding between thermals, birds tend not to resort to active flight. If thermal activity is weak and birds must flap to gain the next thermal, wingbeats are quick, perfunctory, and stiff throughout their length — more choppy than quick and snappy (Sharp-shinned), loose and batty (Red-shouldered), or labored, fluid, and shallow (Red-tailed).

1. A soaring adult Broad-winged Hawk. Size and shape (particularly candle-flame wing shape) are almost idiosyncratic. The black-trimmed, mostly clean and unpatterned underwings and single visible broad white tail band are easy to see. You say you see two bands? Prop the book up and stand ten feet away. Look again. Texas, September NH

2. A lightly marked juvenile Broad-winged Hawk. Note the semblance of a belly band. Also note the paring-knife configuration of the wing in this gliding bird — curved along the leading edge, but straight-cut along the trailing edge. Cape May, October KK

3. A darker juvenile Broad-winged Hawk in a glide showing lots of streaking on the sides of the throat, sides of the breast, and across the belly (but note the unmarked center of the chest). The relatively unmarked and unpatterned underwings are typical. Cape May, October KK

4. Another darker juvenile showing the classic, clean-lined, candle-flame-shaped wings that are so characteristic of this species. Cape May, October CS

5. An adult Broad-winged (showing somewhat ruddier wing linings than is common) in a fast glide between kettles. Wings, in this configuration, are commonly angled down. Note, again, relatively unmarked underwings and, in this case, two apparent broad white tail bands. Okay, maybe just one and a half. Inner band is partially hidden (which is typical). Cape May, October CS

6. A classic Broad-winged Hawk stream between thermals. The paring-knife configuration of the wings (curved along the leading edge, straight-cut along the trailing edge) is manifest, but note, too, the variation from bird to bird. On several adult birds you can also see a single white band on the tail. Texas, September NH

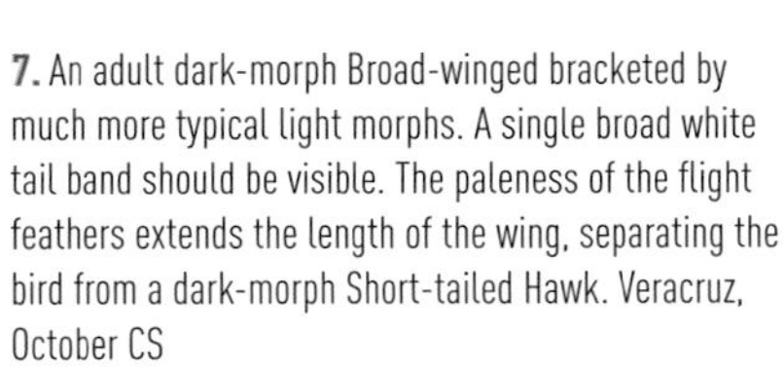

7. An adult dark-morph Broad-winged bracketed by much more typical light morphs. A single broad white tail band should be visible. The paleness of the flight feathers extends the length of the wing, separating the bird from a dark-morph Short-tailed Hawk. Veracruz, October CS

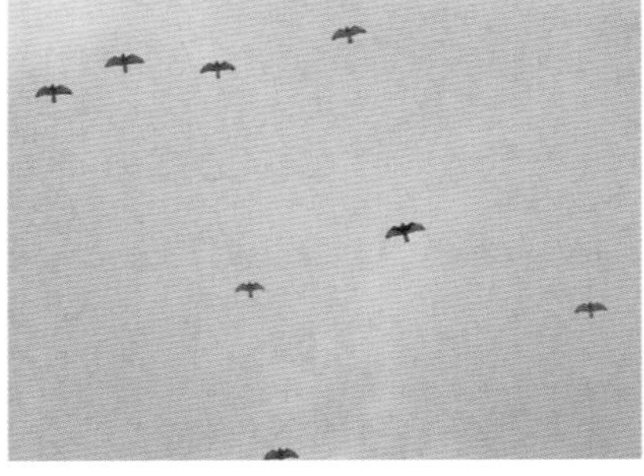

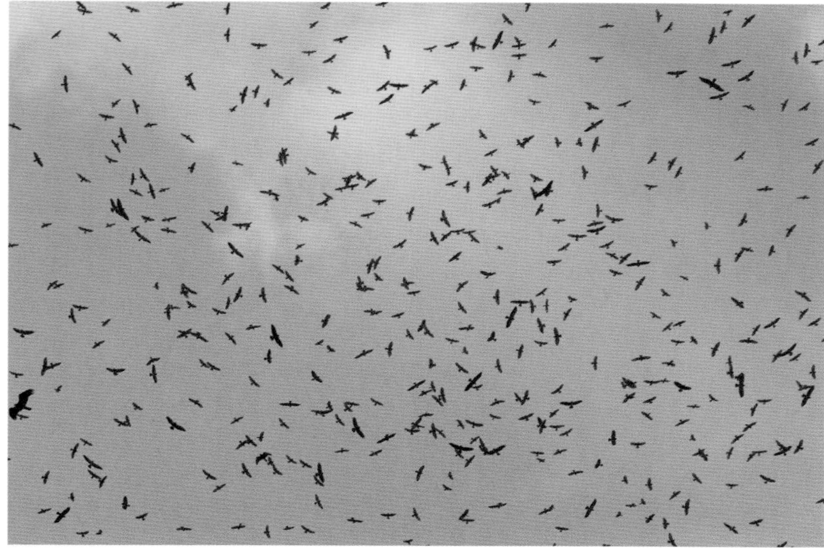

8. A hawk watcher's dream come true: Broad-winged Hawks soaring over Veracruz, Mexico, in October. Picking out the Turkey Vulture should be easy; the several Swainson's Hawks in the mix may take a little longer to find. CS

9. A dark-morph juvenile Broad-winged Hawk. While dark morphs are uncommon in the West and exceedingly rare in the East, the shape is all Broad-winged Hawk. Cape May, October MC

10. Broad-winged Hawks in late spring; juveniles and a single adult. Note the molt in the middle primaries; this disrupts the normal candle-flame shape of the wing. Beware confusing this molt pattern (with light coming through the wings because of missing feathers) with the crescents seen in Red-shouldered Hawk. Always remember that wing shape and appearance can differ when birds are molting (typically in late spring and summer). Cape May, June CS

Swainson's Hawk

The Swainson's Hawk is a handsome and common western raptor (in fact, in summer in parts of the prairies, Swainson's, not Red-tailed Hawk, constitutes the default buteo). Open grasslands and arid and semiarid regions are the bird's habitats of choice. In North America its principal strongholds are the prairies and deserts of the plains states and intermountain West, but some birds range north to the Yukon and eastern Alaska, west to the San Joaquin Valley of California, and east to Manitoba, Minnesota, and Missouri.

The Swainson's Hawk is a champion among long-distance migrants. En masse and in the kettling manner of Broad-winged Hawks, virtually the entire North American population of Swainson's Hawks relocates to the pampas of South America. Seasonal autumn counts in excess of 1.2 million have been tallied at Veracruz, Mexico. In Argentina, during the North American winter, the birds forage communally and opportunistically on grasshoppers and crickets. The preferred hunting technique is simply walking or running up to insects and grabbing them with bills or talons. These raptors also deftly capture dragonflies on the wing.

During the nesting season, Swainson's Hawks take the energetic high road and target larger prey — principally ground squirrels (although adults and young birds consume a variety of both warm- and cold-blooded prey). Versatile hunters, Swainson's Hawks perch-hunt and soar to locate prey and are also adept at coursing low in the manner of harriers.

While the Swainson's is a westerner, it has a well-established pattern of vagrancy. The bird is annual in Braddock Bay, New York, and in Cape May, New Jersey (where the single-season record is 13 birds). There is a small wintering population (made up mostly of juvenile birds) in southern Florida, as well as a few birds in southern Texas and central California. Hawk watchers everywhere should become familiar with this species and consider it as a possibility anytime a raptor is sighted that seems to have an anomalous array of raptor traits.

IDENTIFICATION. Swainson's Hawk is a large, slim-winged, long-tailed raptor that often recalls a harrier, or even a falcon or kite.

The backs of both adults and juveniles are dark except for a light patch at the base of the tail (almost, but not quite, a rump patch). Adult light-morph birds are uniformly brownish gray above; juveniles and subadults are brown to dark brown and variously mottled. Below, in light-morph birds, underwing linings and body are creamy to tawny. The adult has a chestnut bib that covers the chest but leaves a white throat and face.

In dark-morph Swainson's, the body and underwing linings range from rufous to dark chocolate (but almost always with a hint of rufous infused in the feathers). The undertail coverts, serving as a backdrop to the folded feet of soaring birds, are pale. Many dark-morph birds also retain pale patches on the chin or throat.

The streaked and spotted plumages of juveniles (both light and dark morph) generally mimic the pattern of adults, but variations are infinite. In all but the darkest dark-morph juveniles, the heaviest streaking is along the sides of the face, bleeding down to the chest.

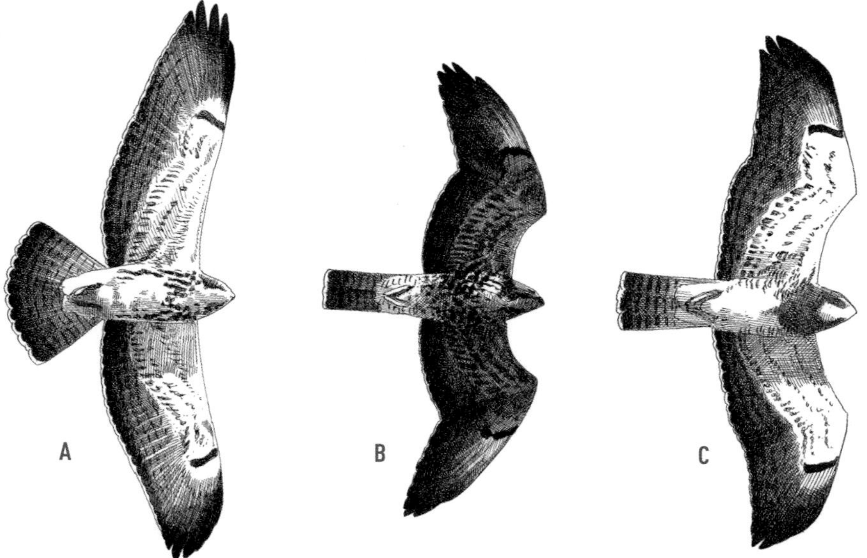

Swainson's Hawk, underside. Light juvenile, soaring **(A)**. Intermediate juvenile gliding **(B)**. Light adult gliding **(C)**. Slender, graceful, and long winged. Wings smoothly curved and pointed; gliding shape **(B, C)** is very distinctive. All plumages have dark flight feathers that are darker toward tip, contrasting in light morph with paler coverts and body. Also note pale spot at base of outer primaries adjacent to dark comma. All show finely barred gray-brown tail. Juveniles are white to buffy below, with streaking concentrated on breast and flanks; darker individuals **(B)** have heavy black spotting across belly and on wing coverts. Adults vary, ranging from typical light morph **(C)**, with rusty brown breast band and fine barring on flanks, to heavily barred rusty on entire underside to entirely blackish, but always with pale undertail coverts (see p. 17 **(D)**).

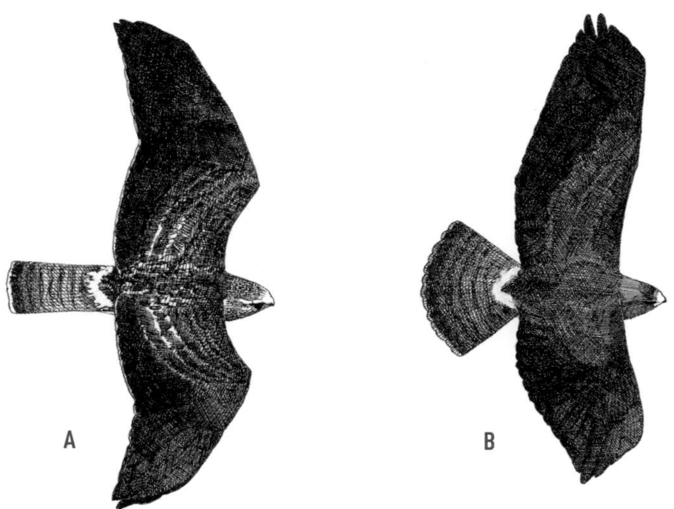

Swainson's Hawk, upperside. Juvenile **(A)**. Adult **(B)**. Both show faint pattern of light coverts and darker flight feathers (mirroring underside); on some individuals this pattern is visible at a great distance. Adult is smooth dark brown; immature has white rump band and white spotting and edging on coverts and scapulars. Some juveniles are strikingly white headed, particularly in spring and early summer when feathers are worn.

In all plumages, the flight feathers — the entire trailing edge of the wing — are dark. On light-morph adults they contrast wonderfully with the paler underwing coverts. On dark-morph adults, the entire underwing (and body) will look dark, but the flight feathers will still look darker than on other dark-bodied buteos.

Juvenile Swainson's Hawk (center) with adult (left) and juvenile Broad-wingeds. Note much darker flight feathers, curved and tapered wings.

On juvenile birds, the outer flight feathers may be paler, with the result that the trailing edge of the wing will appear darkest along the secondaries. Sometimes, too, on extremely pale juveniles (and at close range), the trailing edge of the wing will seem dark only along the border, not unlike that of many other buteos.

More commonly, though, the primaries and secondaries appear smoky dark throughout and the contrast between the pale wing lining and trailing edge is enhanced by distance. The farther the bird, the greater the contrast, so that even paler-winged individuals show a contrast between a pale underwing and darker trailing edge of the wing.

Caution: Swainson's Hawk is not the only North American raptor with an inverse shading pattern — light leading edge, dark trailing edge — to the wing. Adult and juvenile Mississippi Kites have dark underwings and, in a soar, their long tapered wings and similar relative proportions recall Swainson's Hawk. Juvenile and, to a lesser de-

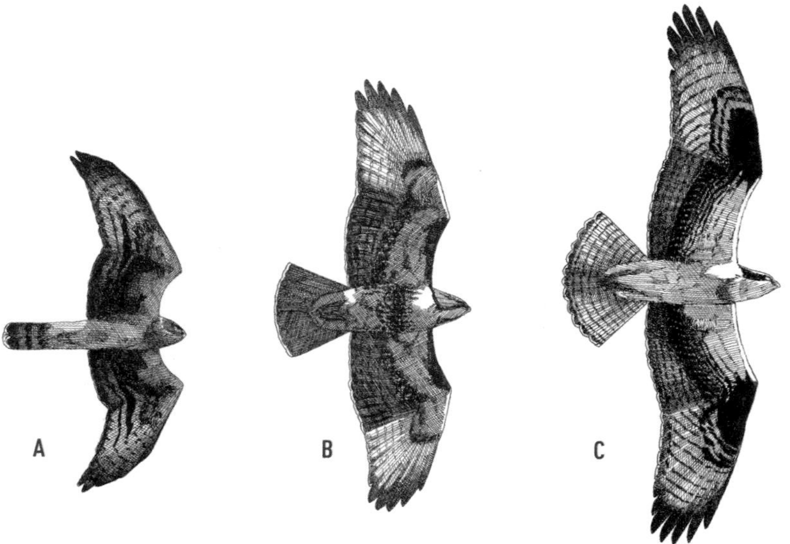

Juveniles gliding. Northern Harrier **(A)**. Red-tailed Hawk (for comparison of size and shape) **(B)**. Osprey **(C)**. Compare to p. 57 top.

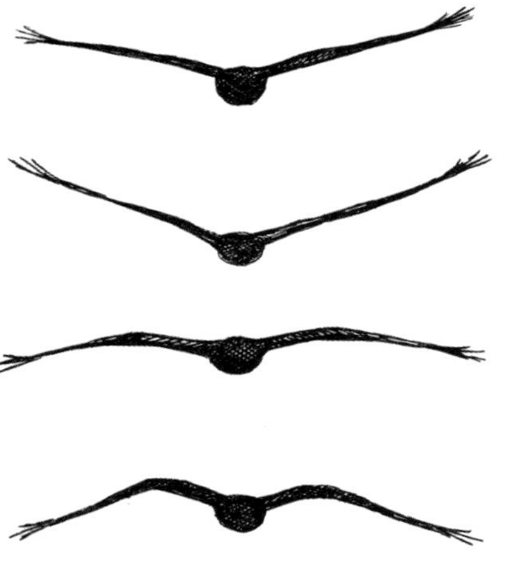

Swainson's Hawk

Swainson's Hawk soars with wings raised in a dihedral; it glides with wings slightly or strongly arched like Osprey.

gree, adult female Northern Harriers and Ospreys also have dark secondaries and pale underwing linings. These species have characteristics that clearly distinguish them from Swainson's, but hawk watchers assuming that a dark trailing edge constitutes an identification quick fix are bound to find a lot of Swainson's Hawks where they are not.

Two other geographically restricted buteo species are easily confused with Swainson's. These are the Short-tailed Hawk of Florida and the White-tailed Hawk of coastal Texas. Both species have a darkly shaded trailing edge of the wing contrasting with white underwing linings, and both have a light-and-dark plumage pattern. Refer to the species accounts of Short-tailed and White-tailed for a discussion relating to separation from Swainson's Hawk.

IN FLIGHT. In a full soar, the wing of a Swainson's Hawk is long, narrow, tapered, even pointed (particularly on juveniles), resembling a long, thin candlestick. It is relatively clean limbed — no overt bumps or bulges — particularly along the leading edge. Broken into component parts, the wing appears both long armed *and* (particularly) long handed — an unusual but not unique combination. Among buteos, Ferruginous Hawk and White-tailed Hawk also share this wing configuration.

The wings are commonly held at right angles to the body but may also be thrust slightly forward. When the tail is fanned, the bird can bear a surprising likeness to a large soaring falcon or Mississippi Kite, another long-armed, long-handed bird.

The wings are also held in a *pronounced dihedral,* with arms stiffly uplifted and hands curved gently up. Swainson's is a light bird with a light wing-loading, and as such it is very buoyant, somewhat tipsy in flight. Wingbeats are slow, stiff, and fairly deep — somewhat harrier-like. And, like harriers and Turkey Vultures, Swainson's Hawks rock in flight like a tightrope walker trying to maintain balance on the high wire.

A gliding Swainson's Hawk holds its wings in a unique configuration. With wrists thrust well forward (projecting beyond the head), the wing configuration combines the dihedral of a harrier and the crooked-wing configuration of an Osprey. It appears as if the bird is resting its weight on the palms of its hands.

In low, hunting flight or viewed head-on, Swainson's can look deceptively harrier-like. Seen in silhouette from below, it is more suggestive of a thin and tapered-winged Rough-legged Hawk.

Rule of thumb: If you initially identify a hawk as a harrier but if, as it approaches, it increasingly takes on the physical attributes of a buteo, you are likely looking at a Swainson's Hawk.

1. A soaring light-morph adult Swainson's Hawk. Classic plumage, classic shape. Arizona, July NH

2. A juvenile dark-morph Swainson's Hawk with a light-morph adult behind. On both birds, note the long candlestick-shaped wings and the dark flight feathers. Arizona, September NH

3. A soaring juvenile light-morph Swainson's Hawk, showing the same basic pattern as adult: light underparts and underwing, dark hood, and dark trailing edge to the wing. But juveniles are buffy to tawny with various amounts of spotting and streaking on the underparts and wing linings. Note, in particular, the exceedingly long, slender, pointy wings. Arizona, September NH

4. An adult dark-morph Swainson's Hawk. Note the combination of dark body and all-dark underwing, but also the characteristic pale undertail coverts. Arizona, October NH

5. Another dark-morph adult Swainson's Hawk, slightly more rufous and richly patterned than #4. Swainson's may show the greatest plumage variation of any North American hawk. Arizona, October NH

6. A soaring light-morph Swainson's Hawk from above, showing this species' pronounced dihedral. Note also the usual light base to the tail. Utah, May JL

7. An intermediate-morph adult Swainson's Hawk in a shallow glide. Intermediates always have rufous tones. In this configuration, note how the bird's long wings and evenly tapered shape seem undiminished. Arizona, September NH

8. A light-morph juvenile Swainson's Hawk. This soaring bird shows the characteristic Swainson's dihedral. Arizona, September NH

9. A juvenile light-morph Swainson's Hawk in a glide, showing the classic "M-shaped" glide profile that recalls, somewhat, an Osprey. Utah, August JL

10. An adult intermediate-morph Swainson's Hawk in a glide. Not as steep or fast a glide as in #9, but very characteristic of long-winged Swainson's — and unlike almost every other buteo. Nevada, September JL

11. They travel in packs when on migration: a kettle of Swainson's Hawks over Cardel, in Veracruz, Mexico. Light, dark, and intermediate morphs can be picked out in this photo, although most, if not all, appear to be adults. October CS

Rough-legged Hawk

This is an Arctic-breeding buteo, feathered to the foot, hence the name. Most raptors have bare tarsi. Unlike other North American buteos, the Rough-legged Hawk is not limited to the New World. Its distribution is circumpolar; breeding populations are found above the tree line across northern North America, Greenland, Europe, and Asia.

Like their Arctic neighbors, the Gyrfalcon and the Peregrine Falcon, Rough-legged Hawks are cliff nesters. Tundra without cliffs (or elevated manmade structures that substitute for cliffs) is devoid of breeding Rough-leggeds.

In winter, the entire population shifts south, occupying most of the United States and adjacent portions of southern Canada and northern Mexico. Across the southeastern United States, it is a rare, not regular, winter visitor. In the West, the range of wintering birds straddles the United States–Mexico border.

In winter, the preferred hunting habitats are low and wet marsh, agricultural areas, pastures, and grasslands. While the bird does hunt from a perch, often sitting higher and on springier branches than other buteos, the Rough-legged is primarily an aerial predator, at home where perches are few and where its aerial finesse outclasses the competition.

Hunting Rough-leggeds glide toward an auspicious-looking site and hover—sometimes, in a good breeze, moving just the wingtips; sometimes using deeper, more thrusting flaps. The bird's head sweeps from side to side as it searches the ground below for small rodents, its favored prey. If the search proves futile, the bird glides, or more often uses powered flight, to move to the next potential site. If prey is sighted (and/or heard) the bird drops or parachutes to the ground, legs extended.

Rough-legged hawks usually hunt from heights of between 40 and 150 feet. But Rough-leggeds may also beat the bushes, particularly in hilly country, in the low-flying fashion of a harrier.

Fall migration begins in mid- to late September, peaks in late October or early November, and continues into December (even January). Heavy snowfall may prompt birds to resume their migration even as late as February. In spring, some Rough-leggeds begin moving north as early as mid-February as birds adjust their winter territories to take advantage of the spring thaw. Migration begins in earnest in March, peaks in April, and continues into early May. The species is irruptive (winter concentrations shift in response to the abundance or lack of prey).

The Rough-legged Hawk is one of the least common migrants at eastern hawk-watch sites, and numbers are diminishing. Hawk-watch sites on the north side of the Great Lakes (in fall) and south side (in spring) constitute the best locations to study numbers of these northern birds. Key locations include Duluth, Minnesota, in the fall and Whitefish Point, Michigan, and the south shore of Lake Ontario in the spring, when scores of birds may be seen in a day.

IDENTIFICATION. The Rough-legged Hawk is a large, lanky, angular buteo. In all plumages it is boldly patterned, a characteristic that sets it apart from many buteos. These plumage differences relate to age, sex, and two recognized color morphs. In

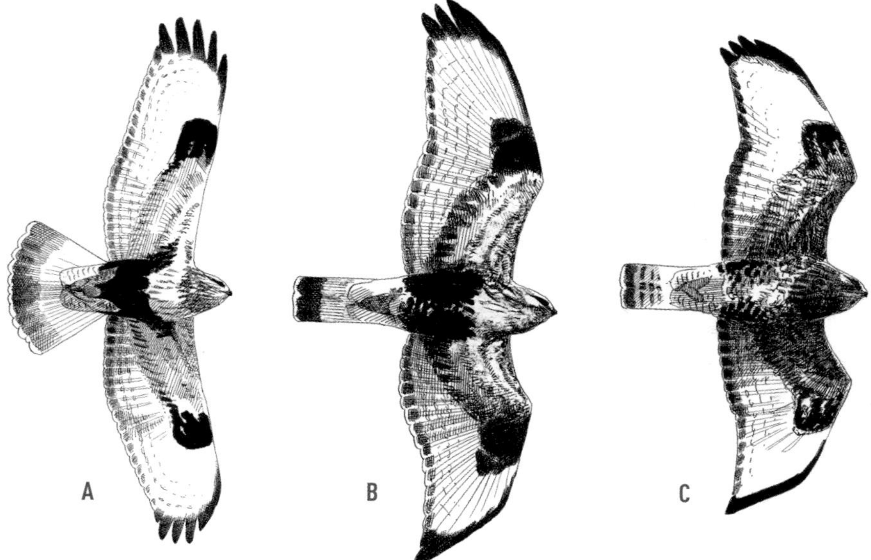

Rough-legged Hawk, underside. Juvenile **(A)**. Adult female **(B)**. Adult male **(C)**. Long winged, lanky, and angular; long tailed; head large and round. All plumages show blackish carpal patch and silvery flight feathers; most show blackish belly patch, but adult males **(C)** typically have breast darker than belly. Juveniles **(A)** differ from adults in having little or no streaking or spotting below; solid blackish belly and carpal patches; and pale gray, not blackish, tips on primaries, secondaries, and tail feathers. For dark morph see p. 17 **(C)**.

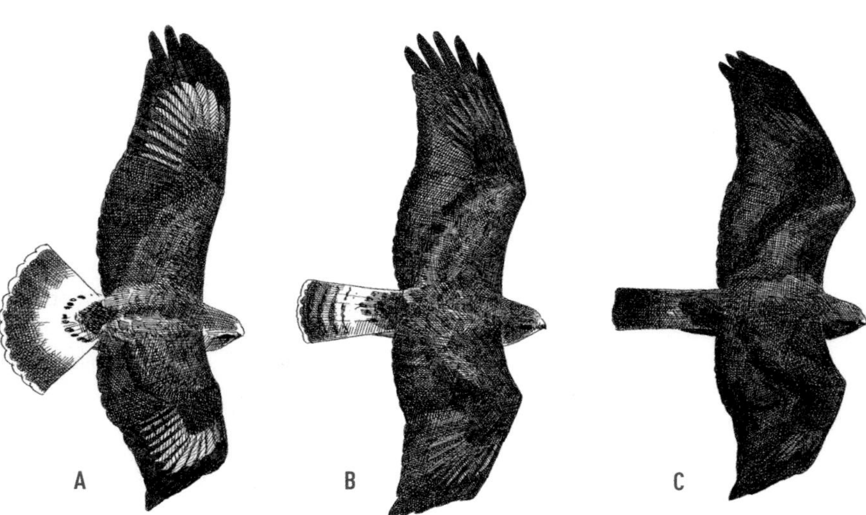

Rough-legged Hawk, upperside. Juvenile light extreme **(A)**. Typical adult **(B)**. Dark extreme **(C)**. All have dark greater coverts, contrasting with paler back and lesser coverts. All lighter individuals show white base of tail. Adults have barred tails. Juveniles may show some white in primaries, but variation is considerable and complex.

the East (north of Florida) a dark-morph buteo is much more likely to be a Rough-legged Hawk than any other species. In the West, light-morph Rough-leggeds greatly outnumber dark morphs but in the East, at some hawk watches and winter sites, dark-morph birds may be nearly as common as light morphs.

Distinguishing light-morph Rough-leggeds from other buteo species is pretty straightforward. Distinguishing dark-morph Rough-leggeds from other dark-morph buteos is more challenging.

In light morphs, adults and juveniles are dark above with contrasting paler heads, pale wing patches bleeding through the hand (absent in adults), and a white tail tipped with a broad dark subterminal band. The overall color pattern of most adult males is black, gray, and white; adult females and juveniles are dark brown, tawny to buff, and white.

Underparts are pale, creamy white (adult males) or tawny to buff (females and juveniles). *In all light-morph individuals, wings boast large dark carpal patches.* These patches are most prominent on juvenile birds, whose underwing linings are typically pale and indistinctly marked. The underwing linings of adults are more heavily marked, a backdrop that is creased by a pale median bar — a pale support strut running the length of the darker wing lining from the body to the carpal patch.

White tails are trimmed with *a prominent dark subterminal band on the dorsal side.* At close range and from below, tail bands appear diffuse on juveniles; prominent, single- or double-banded and contrasting on females; and sharply defined and multibanded on adult males. Viewed at a great distance, these subtleties disappear, and all birds show a dark subterminal band that appears solid and well defined.

Females and (particularly) juveniles have a dark belly — a belly band that is wider, darker, and more prominent than the band of Red-tailed Hawks. The belly of juveniles can appear black!

In an adult male, the entire body is darkly mottled, almost calico or checkerboard patterned, except for a pale swath between the bib and belly — an inverse necklace where the color has been bleached out of the bird. In some adult males, the patterning on the belly is limited to the flanks or absent. Such birds have a bibbed more than a hooded appearance.

The upperparts, body, and wing linings of adult male dark-morph Rough-leggeds are most commonly *black;* females and juveniles are dark chocolate. White on the upperwings and uppertail may be limited and is often absent in adults.

From below, the flight feathers of dark- and light-morph Rough-legged Hawks are pale, silvery onto white, with some barring on the secondaries and inner primaries. Barring is absent on the outer primaries. The resulting translucent patch sandwiched between the dark carpal patch and the dark wingtip is particularly prominent when birds are backlit (and most prominent on juveniles).

Confused? If you want to simplify the process, just remember this: all light-morph Rough-legged Hawks show a pale head and a broad, dark band near the tip of their white tail. Below, birds are boldly marked. Most have a dark belly and a dark-banded white tail. All have dark carpal patches and show a pale underwing patch near the wingtips (which bleeds through to the upperwing on juvenile birds).

The tail of most dark-morph birds has the bold, dark terminal band — a mark not shared by most other raptors. In the East, the only bird with a similar tail pattern is the immature Golden Eagle. In the West, adult Harlan's Hawk is a near look-

Rough-legged Hawk

Rough-legged Hawk soars with wings angled slightly up in a graceful dihedral; it glides with wings raised and angled in a modifed dihedral.

alike. Shape is useful when distinguishing these species, but the dark-morph Rough-leggeds' prominent white underwing patch is diagnostic.

IN FLIGHT. The eye-catching array of Rough-legged plumages masks an important point: the shape of Rough-legged Hawk is *very* different from that of other buteos. The body is robust, the head prominent, the bill disproportionately small. The tail is buteo-wide but longer than your stock model buteo.

The wings follow suit — very long and uniformly broad along their length. The lines are clean, trim, and straight. The blunt-tipped hands are very modestly tapered, imparting a sense of grace to the wing shape.

In sum, Rough-legged is a long, lanky, yet sturdy-looking buteo. If the bird were to play professional sports, it would look at home playing center on a basketball court.

From below, in a full soar, the wings are thrust slightly ahead of the bird. In a shallow glide, the arm remains thrust forward, while the hand curves gently back. In a more severe glide, the arm is fixed at a right angle to the body, but the very long hand sharpens to a point — the leading edge curves back and the trailing edge angles sharply back.

Head-on, the bird soars and glides with a pronounced dihedral, with wings that jut up sharply from the body at the shoulder and flatten out at the wrist. Though many buteos fly with a dihedral, the Rough-legged alone shows this exaggerated up-and-out, hunch-shouldered configuration. It is more buoyant than Red-tailed and slightly more tippy or unsteady, too, but not as rocking or unbalanced as harrier (or Swainson's Hawk).

In level flight, Rough-leggeds retain their long-winged appearance, although the taper becomes acute. The wingbeats are unhurried, even, and deep and project a sense of rowing. Each wingbeat is methodical and precise, making the flight purposeful and direct. It recalls some long-limbed Viking oarsman pulling to the beat of an unheard drum.

Of all the members of the buteo clan, Rough-legged Hawks seem most adept at powered flight. Hovering more often than kiting, hunting birds shift locations by powering over to a more promising site on steady wingbeats, gliding infrequently if at all. They readily use direct, active flight to cross open water. Commonly, this is the first buteo to appear in the morning (while other buteos are waiting for thermals to perk) and the last to be recorded in the afternoon — long after the day's thermals, and other buteos, have left the sky.

1. A juvenile light-morph Rough-legged Hawk showing classic plumage points: dark carpal patches, broad dark belly band, broadly but not crisply banded white tail. Note, too, that the dark trailing edge of the wing is fairly pale and diffuse. Utah, October JL

2. An adult light-morph female showing dark belly and paler chest, and for the most part a single prominent dark tail band (some birds show a second, narrower inner band). This soaring bird shows the classic, very long planklike wings. Note the darker and more distinct border to the trailing edge of the wing. The border on juveniles is paler and more diffuse. Utah, October JL

3. An adult light-morph female Rough-legged Hawk soaring or riding an updraft. This bird shows, faintly, the semblance of a second inner tail band. Utah, October JL

4. An adult male light-morph Rough-legged Hawk, wearing a bib but no belly band. Note the multiple bands on the tail, which are characteristic of adult males. Utah, October JL

5. An adult male light-morph Rough-legged Hawk in a glide showing a prominent bib contrasting with a pale belly and multiple tail bands. Note the wing shape, showing long swept-back hands. Alaska, July TS

6. An adult male Rough-legged Hawk. This is the archetypal, lodge-this-search-image-in-your-mind shape of a Rough-legged gliding overhead. Note, again, the long tapering hands and the fairly long but broad tail. Note also the very dark flank patch, which is typical of light-morph adult males and, again, the multibanded tail. Alaska, April JL

7. A soaring adult Rough-legged Hawk, dorsal view. Note the modified dihedral (wings that lift up along the arm and flatten along the long hands). Dark band on the white tail is diffuse. Pale head is more typical of juveniles, as are the white patches bleeding through the upperwing that recall some well-marked juvenile Golden Eagles. Note also that the white tail on a Rough-legged hawk appears just as brilliant as the tail of juvenile Golden Eagle. Utah, October JL

8. An adult female dark-morph Rough-legged Hawk. Note first the classic broad, planklike wings. You can tell it is an adult because of the well-defined single dark band on the tail and very dark trailing edge of the wing. Note that the black carpal patches are still apparent despite the dark wing lining. Utah, October JL

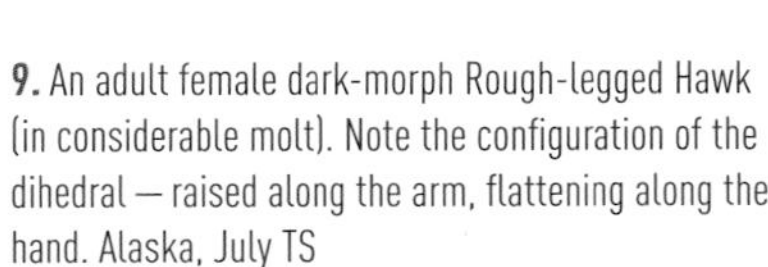

9. An adult female dark-morph Rough-legged Hawk (in considerable molt). Note the configuration of the dihedral — raised along the arm, flattening along the hand. Alaska, July TS

10. A dark-morph Rough-legged Hawk in a glide. The contrast on the underwing is increased because this bird was photographed over snow, making the whites brighter, but such is often the case with this northern buteo. Alaska, April JL

Ferruginous Hawk

If the Ferruginous Hawk were an Old World raptor, it would almost certainly be classified as an eagle. In size (and perhaps prowess), the bird stands midway between the buteos and the *Aquila* eagles. North America's largest buteo, the Ferrug manifests the hunting skills (and many of the techniques) of a Golden Eagle.

Ferruginous Hawk is a westerner, a bird of dry, open country—sage flats, short-grass prairie, and desert. Its breeding range begins where even the rumor of eastern forests doesn't extend: the prairie regions of western South Dakota, Nebraska, Kansas, Oklahoma, and Texas. In winter, birds expand their range east in the states mentioned, west to California, and south into northern Mexico.

Small mammals are the bird's preferred prey—ground squirrels, pocket gophers, and prairie dogs in particular. Jackrabbits, prairie chickens, pheasants, and even sage grouse are not overlooked, particularly in winter, when in many places a period of hibernation puts favored prey species out of reach.

The Ferruginous Hawk is an active hunter, hovering like a Rough-legged Hawk, kiting like a Red-tailed, and quartering high over the plains, with its wings raised and its head turned down as it studies each promising movement below.

During warmer months, when rodent prey is active, the bird frequently hunts by cruising low in the manner of a Golden Eagle, another hill-country hunter. The long wide wings are drawn and tempered by the wind to hunting points and the heavy, powerful, tireless wingbeats and long sweeping glides make a Ferruginous easy to mistake for an eagle—an eagle in buteo clothing.

Ferruginous Hawks perch-hunt, too, commonly from a slightly elevated patch of ground. Only harriers spend as much time sitting on open ground as this bird does. Other buteos seem out of place sitting in open terrain, but the Ferruginous Hawk's long, low profile coupled with its alert, erect stance and elevated head make it appear very much at home on the ground. These birds often prefer ground perches, even when trees or some suitably elevated perch is nearby.

In the core of its breeding range, the bird is a permanent resident. However, as the winter redistribution of birds into northern Mexico attests (and the incidence of vagrants in the Midwest, even Florida, supports), some Ferruginous Hawks, including all those located in the northern half of the bird's breeding range, engage in modest to fairly lengthy migration (hence, its inclusion in this chapter).

IDENTIFICATION. The Ferruginous Hawk is a large, robust-bodied, yet lanky buteo wrapped in striking plumage. The head is wide and wedge-shaped, the forehead sloping, the bill large and eaglelike—a doorstop with regal overtones. The body is heavy and barrel chested. In combination with the bird's prominent head, Ferruginous Hawk has a top-heavy (in flight, front-heavy) appearance.

Two color morphs are recognized: light and dark. The dark upperparts (head, back, upperwing coverts) of light-morph birds are infused with bright, rusty brown—rusty, as in oxidized iron, as in ferrous oxide, hence *ferruginous.* Some birds have a pale gray face with a contrasting darker eye line. Most birds are dark capped and appear dark headed at a distance.

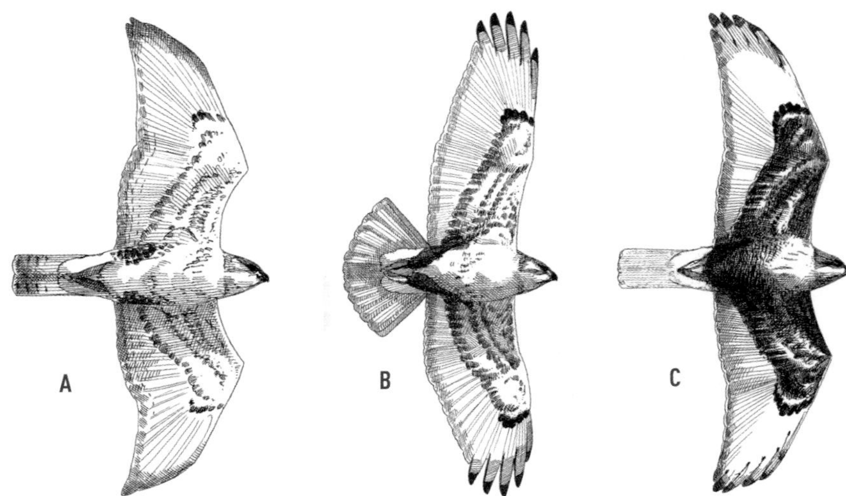

Ferruginous Hawk, underside. Juvenile **(A)**. Typical adult **(B)**. Heavily marked adult **(C)**. Large and long winged with pointed wingtips, broad armed and small handed. Wings can appear very pointed in glide, especially on juvenile **(A)**. Head and bill larger than on Rough-legged. Body normally strikingly white; dark birds such as the one shown in **(C)** are very uncommon. Underwing sparsely spotted; may show dark patagium but only when entire underwing is heavily marked as in **(C)**; all have dark comma and silvery white flight feathers with little or no barring and gray trailing edge. Adult has rusty blotching on underwing, rusty leggings, pale gray and rufous tail. Much variation between **(B)** and **(C)**. Juvenile **(A)** has blackish spotting on underwing and lower flanks; less-defined gray tips on flight feathers; and faint barring on secondaries and tail. Also see dark morph on p. 16 **(B)**.

Ferruginous Hawk, upperside. Juvenile **(A)**. Adult **(B)**. Adult is bright rusty on back and wing coverts. Flight feathers gray, with white showing on inner webs of primaries. Tail gray and rufous with white base and white inner webs; generally looks gray-brown when closed and exposes white inner webs when fanned. Head varies from whitish to dark gray. Juvenile is brown with white spotting and extensive white in primaries and secondaries. All ages show dark greater coverts and pale base of primary coverts.

A juvenile Ferruginous Hawk in typical posture, perched on the ground.

Prominent on the rusty upperwings are white patches on the hand — a mark that recalls the upperwing pattern of many juvenile and subadult Golden Eagles. Prominent in both adults and juveniles, the white wing patches above and below are visible from a tremendous distance. Tails, too, are prominent and pale — white near the body, though washed (sometimes streaked) with rufous nearer the tip.

On most individuals, underparts are white, variously marked with an array of rufous-brown patterning, and the chest is conspicuously white with little or no patterning.

In very light individuals, the patterning may be little more than rufous spotting on the underwing linings and pale barring along the flanks and legs. In darker individuals, almost the entire underwing lining is dark rufous (a trait that will recall the dark wing struts of Prairie Falcon), and flanks and legs (even the belly!) are darkly barred.

Most individuals show prominent commas at the wrist, and a patagium that is more patterned or more darkly shaded than the balance of the underwing linings. Most individuals have a pale, unbanded tail that may be mostly rufous above but appear white below except for a blush of color near the tip. The dark rufous legs, folded against pale undertail coverts, form a distinct V.

Juvenile light-morph birds are colder brown above, with little or no rufous highlights, and less heavily marked below. They show very little or no barring on the legs — and *no* telltale V. The tail is finely barred, with bars darker and more clustered toward the tip. At a distance, this may suggest a single dark subterminal band. Primaries and secondaries are lightly barred, unlike those of adults.

The less common dark-morph Ferruginous (outnumbered by the light morph 20 to 1) are dark reddish brown or blackish brown above (with just a hint of warm tones). The underparts — body and underwing linings — seem cast from cinnamon-dusted chocolate. The head is grayer and colder toned. A darker comma at the wrist is usually separated from the underwing lining by a pale comma.

Because the undertail coverts are darkly banded, the legs form no telltale V. The tail is steel gray or washed with rufous above and white below.

Dark-morph juveniles are uniformly sooty or blackish brown above, showing far less rufous than adults. The tail is barred and overall much darker than that of adults. The characteristic wing pattern is less distinctive, but present. From below at a distance, the undertail may show a dark subterminal band.

IN FLIGHT. This is a big, wedge-headed, broad-chested, long-winged, long-tailed bird. In shape, it combines the robustness of Rough-legged Hawk with the slender elegance of Swainson's Hawk. In the light morph, the conspicuously white barrel-shaped chest stands out (even in flight). This feature, coupled with the bird's wide, projecting, wedge-shaped head gives the bird a front-heavy appearance.

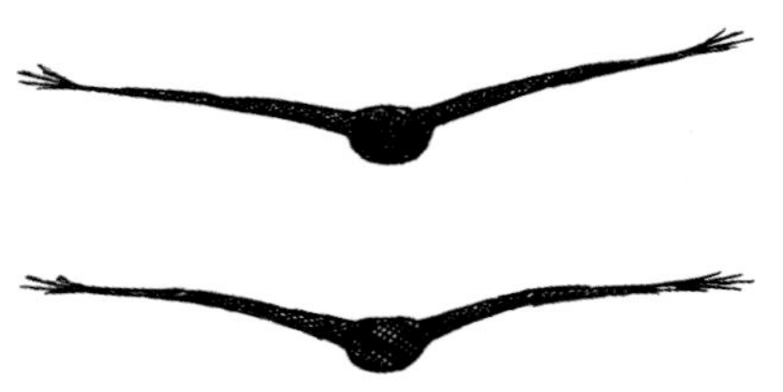

Ferruginous Hawk

Ferruginous Hawk soars with wings raised stiffly in a moderate dihedral and glides with wings raised and angled in a modified dihedral.

Adult Rough-legged Hawk (left) with adult Ferruginous (right). Note differences in proportions: the Rough-legged has wings broader in the hand, shorter head, slighter body.

From a distance, light-morph birds are overall much paler than the buteo norm. The inky black primary tips, so common in most buteos, are inconspicuous to invisible. The white wing patches and tail stand out as three points of light at tremendous distances. On some individuals, the dark wing struts, contrasting with the pale flight feathers and chest, are conspicuous at distances that render the dark V leggings indistinguishable.

In a full soar, the wings are exceedingly long, broad, and eaglelike; the lines are smooth and clean, not lumpy or bulging like those of a Red-tailed Hawk. The leading edge is straight-cut; the trailing edge curves gently, narrowing into a tapered tip where the arm meets the body — imparting to the trailing edge of the wing a distinct, curvaceous, S-shaped configuration.

The arm is long and wide, while the hand is short, tapered, and noticeably thinner. (Red-taileds have evenly proportioned wings with broad hands; Swainson's Hawks and Rough-legged Hawks are long armed and long handed.) The tail, too, is long and, even when the bird is soaring, often closed.

Like many western buteos, the Ferruginous Hawk soars with a dihedral. The entire wing is uplifted, stiff, or very gently curved along its length — not bowed, not jutting at the shoulders or uplifted along the hands — symmetrically lifted along its length (as if the bird were invoking a prairie spirit).

In a glide, the dihedral becomes more acute along the arm, and the wingtips draw back to points, imparting to this eaglelike buteo the lines of a falcon or a kite.

The wingbeat is more deliberate but more fluid and elastic than that of a Red-tailed. It recalls a heavy-handed Swainson's Hawk or a harrier with the lumbering, almost mechanical grace of a Golden Eagle. Contrary to some accounts, the Ferruginous Hawk hovers frequently with a style somewhat recalling a hovering Osprey — deep wingbeats with a great deal of wrist movement. In cruising flight, it frequently hugs the contours of hills (like a hunting eagle) and in a soar moves its head side to side in the fashion of a Rough-legged Hawk.

An adult Ferruginous Hawk gliding head-on.

1. The ventral view of an adult Ferruginous Hawk, showing the classic long tapered wings, barrel-chested body, dark wing linings, and dark legs contrasting with a white trailing edge on the wings and white tail. This is a fairly heavily marked light morph. Utah, October JL

2. A very lightly marked (and white!) light-morph adult Ferruginous Hawk. Very well represented in this photo are the robust body, wing shape, and whitish tail. Note, too, the doorstop-shaped head. At times when a bird is flying over snow, reflected light can make it look brighter and whiter than usual. Utah, October JL

3. An adult Ferruginous Hawk in a glide. Even slightly tucked, the great length of the tapered wing is apparent. So, too, is the hint of a dihedral. Utah, September JL

4. An adult Ferruginous Hawk banking in a soar. The three points of light — wingtips and tail — are apparent at great distances. Note here the graceful dihedral. Utah, May JL

5. A soaring adult Ferruginous Hawk. This one is classic, showing a barrel chest, acutely tapered hand, dark underwing linings, and whitish tail. Utah, September JL

6. A juvenile light-morph Ferruginous Hawk — a paler individual than in #7, an appearance perhaps magnified by snow cover below. Note the clean white underparts and white throat with dark eye stripe and crown. Utah, November JL

7. A juvenile light-morph Ferruginous Hawk, showing lightly banded tail and just a little more patterning below than normal (including the semblance of a belly band). Notice the straight leading edge of the wings. Utah, October JL

8. A dark-morph adult Ferruginous Hawk, gliding. Dark chocolate body, whitish tail without a dark subterminal band, and white crescents near the outer tip of the underwing coverts (not to mention shape) distinguish it. Utah, May JL

9. A hunting adult Ferruginous Hawk, showing a textbook dihedral. Arizona, December NH

10. A dark-morph adult Ferruginous in a fast glide. The shape is classic Ferruginous: long tapered wings, barrel chest, longish white tail. Clearly evident are the white commas on the underwings bordered by black. The whitish panels near the wingtips? Seen from above, they appear as white wingtips. Add the white tail and you get the classic Ferruginous three points of light. Utah, May JL

11. A gliding adult Ferruginous Hawk, high overhead. The barrel chest, long tapered wings, and closed tail are classically diagnostic. There is a hint of a belly band, but this is no Red-tailed! Arizona, January CS

PUTTING IT TOGETHER: Telling Buteos Apart

At first glance, owing to their distinctive plumage patterns, buteos might seem fundamentally easier to identify than falcons or accipiters. The identification process, however, is complicated by an array of plumages relating to age, color morphs, and subspecies and further complicated by overlapping traits between species and wide variation in the plumage of individual birds within a species.

And that's not all. In North America, there are only three accipiter possibilities to ponder and three or four viable options when it comes to pinning a name to a distant falcon. (Gyrfalcon and Aplomado Falcon are extremely unlikely in most places.) But over much of North America, there are five or six buteo possibilities to consider. Every additional variable complicates the identification process.

Begin by managing variables before you start watching. Observers at most points on the continent can usually eliminate one or two possibilities on the basis of probability and range. Broad-winged Hawks are not likely in Utah. Rough-legged Hawks rarely reach Georgia. Swainson's Hawks are not everyday fare in New Hampshire. Ferruginous Hawks make very few appearances in Ohio (or Pennsylvania, Maryland, Delaware, Quebec . . .).

In short, know the range maps in your field guides.

It is also helpful to bring temporal considerations into the identification process. In Minnesota, a distant buteo in September might be a Broad-winged Hawk. In November, when most Broad-winged Hawks have retired to more temperate areas, other, more seasonally calibrated possibilities (like Red-tailed Hawk or Rough-legged Hawk) are more likely.

After the possibilities are pared down, buteo identification usually comes down to a choice between two or three realistic possibilities (is it a Red-tailed or Red-shouldered?) or may be a simple matter of confirming or refuting the expected occurrence — the default bird.

Say, for example, you are standing atop Mount Tom, Massachusetts, and it is September 14 — it is a very good time to see migrating Broad-winged Hawks. Experienced hawk watchers know this (and information relating to numbers and occurrence at hawk-watch sites across North America is easy to come by). The logical thing, the hawk-watcher thing, would be to look at an approaching distant buteo and see whether the characteristics you are noting are consistent with Broad-winged.

Check to see whether it appears small, compact, and short tailed, with candle-flame-shaped wings (characteristics more easily noted at a distance) before straining to note things such as windows, commas, belly bands, tail bands, patagial marks, and eye lines — characteristics that present themselves more easily as the distance shortens.

Play the top card, the likely card, first. Then, if the bird you see doesn't seem to fit, check whether it seems more consistent with another possibility — which very likely is the case!

As noted earlier, migrating Broad-winged Hawks commonly fly in flocks. On September 14, atop Mount Tom, Massachusetts, if you see one Broad-winged, in very short order you will see many more as additional Broad-wingeds catch up to what is likely the lead bird in a group. A solitary buteo on Mount Tom in September that remains solitary is more likely to be a Red-tailed Hawk, the other likely buteo card in the deck, and a buteo whose very bold plumage pattern (not to mention its penchant for kiting) will soon be apparent.

In the case of very distant soaring birds, plumage characteristics are commonly difficult to make out. Some bold, blatant marks can vault the distance: the red tail on a Red-tailed Hawk, the three points of white on a Ferruginous Hawk, the gleaming white tail on a Rough-legged Hawk. But lighting conditions may be poor and distances are sometimes too great for even marks as prominent as these to be noted.

Motion, the rhythm and pattern of a bird's flight, projects itself across vast distances. How steady a bird is on the wing and the frequency, manner, and rhythm of a bird's flap are qualities that are easily noted. The trick is decoding them.

For example, heavy birds, such as Red-tailed and Ferruginous hawks, soar and glide with little or no wobble. The wingbeats of a Ferruginous are slow, stiff, and pushing — very eaglelike. By comparison, the wingbeats of a Red-tailed are quicker and more abrupt, more fluid and flexing, and fairly shallow.

Swainson's Hawks are buoyant, almost acrobatic in a thermal. Their wingbeats are slow, stiff, fairly deep, and measured — reminiscent of harriers. Both Swainson's and Rough-legged hawks tip unsteadily in a wind, and Rough-leggeds make use of powered flight — they fly using slow, even, deep, purposeful wingbeats, while other buteos are more wedded to a less energetic style, relying upon thermals for lift and long steady glides to gain ground.

Red-shouldereds are high-strung and often surrender to a series of quick, floppy wingbeats even when gliding between thermals or riding a strong supporting updraft. Broad-wingeds, being the smallest of the migrating buteo types, turn in very tight circles. In a kettle, the birds are densely clustered and not widely spaced; in active flight, the wingbeats are hurried, stiff, and choppy.

As a bird approaches, shape is usually the next characteristic to project itself. Sometimes a buteo's way of holding its wings offers a clue. Red-tailed Hawks soar and glide with wings uplifted in a V or an exaggerated U. Ferruginous Hawk, too, soars and glides with its wings held in a V — but one that appears wider, stiffer along its length, and slightly less acute. Rough-legged soars and glides with a dihedral that juts up at the shoulders and flattens out across the hands. Swainson's Hawks soar with an exaggerated V, and in a steep glide, the wings assume a bowed, gull-like, or Ospreylike shape. A Red-shouldered may soar with its wings very slightly uplifted or very slightly drooped, but in a glide the wings are bowed down as if the bird were cupping the air. Broad-winged Hawks soar on flat wings and glide with wings angled stiffly down.

Seen well, even allowing for variations among individuals, the composite shape or silhouette of each of the species discussed in this chapter is as diagnostic as any field mark relating to plumage. *Study the photos and illustrations.* Fix the shape of the birds in your mind and bring these templates to bear in the field. Recognize that the angle of view and the flight attitude of the bird will affect the shape you see.

- The basic Red-tailed is big, broad, lumpy winged, and short tailed.

- Red-shouldered is slighter, trimmer, and more clean-cut than a Red-tailed, more gentrified. Wings reach forward in a soar. Juveniles are plank winged, lanky; adults, paddle winged and compact.

- Broad-winged is small and compact. The wings are flame-shaped in a soar, paring-knife-shaped in a glide. Gliding juveniles, with wings drawn back and narrow notched tails, may remind you of Sharp-shinned Hawks.

- Swainson's Hawk is like a Broad-winged painted by El Greco — a slim, clean-limbed bird with wings that look like long tapered candlesticks.

- Rough-legged Hawk is long limbed and broad winged with a tail to suit. The arm is long, and so is the hand. The wingtips are blunt or sharp-cut.

- Compared to other buteos, Ferruginous Hawk appears distinctly big headed, barrel chested, and front-heavy. To carry this load, the wings are exceedingly wide and long at the arm; to add grace, the hand is acutely tapered and disproportionately small.

At any point in your study of a distant buteo, the glimpse of a key plumage characteristic — such as those listed here — can confirm or refute a tentative identification.

- Red-tailed Hawk: Red tail of most adults. Belly band of most juveniles and many adults, or (conversely) the white chest or white underparts of lightly marked birds. The two-toned upperwing (dark arm, pale hand) of juveniles.

- Red-shouldered Hawk: The pale crescent-moon window that halos the tips of the wing.

- Broad-winged Hawk: Single broad white band bisecting the tail on adults. Comparatively clean, unmarked underwings of both adults and juveniles.

- Swainson's Hawk: Contrastingly dark or shadow-colored trailing edge of the wing. In a glide, it recalls an Osprey.

- Rough-legged Hawk: Dark-banded white tail. White wing patches on juveniles. Dark belly and paler chest. In the East, north of Florida, it is the most likely dark-morph buteo.

- Ferruginous Hawk: Three points of white (wing patches and pale tail).

Finally, when the bird is close enough, you can begin to note less conspicuous, plumage-based characteristics — patagial lines, commas on the wings, banded (or unbanded) outer primaries — the field marks that serve to distinguish very similar species and confirm or refute a tentative identification.

Accipiters

The Artful Dodgers

SPECIES
Sharp-shinned Hawk, *Accipiter striatus*
Cooper's Hawk, *A. cooperi*
Northern Goshawk, *A. gentilis*

Sharp-shinned Hawk

Cooper's Hawk

Northern Goshawk

All three accipiters have similar wing positions in flight. They soar with wings flat or slightly raised and glide with wings pulled in, flat, or slightly bowed. The wings are thicker and have more obvious brushlike fingers at the tips than do falcons' wings.

ccipiters are quick, agile raptors of boreal forests, bayberry thickets, sycamore canyons, and backyard bird feeders. Here, evolution's push and pull has created a group of birds designed to capture other birds (and some mammals) in deep woods, in thick growth, and around corners.

They would be much more appropriately called short-range interceptors.

Accipiter wings are typically short and rounded, adapted for rapid acceleration and tight confines. The tails are long and rudderlike, ideally suited for nimble acrobatics around an obstacle-strewn course—a course no falcon or buteo could follow. If a Sharp-shinned Hawk is released in front of a seemingly impenetrable maze of branches, the bird will melt into the foliage with only a vibrating twig or two to mark its entry point.

Accipiters are short-range specialists. Their forte is the sprint, their ally surprise (not endurance). If they are unsuccessful in taking prey in the first effort or after a short pursuit, they commonly break off the chase.

LEFT: A juvenile Cooper's Hawk. All accipiters have proportionately short wings and long tails that allow exceptional maneuverability. KK

Three species of the genus occur in the United States and Canada. In order of abundance (over most of their ranges), from smallest to largest, these are the Sharp-shinned Hawk, Cooper's Hawk, and Northern Goshawk.

MIGRATION

Sharp-shinned and Cooper's hawks are migratory in the most classic sense. Most members of the populations move south before the leaves have fallen from the trees, following timetables that correspond to the migration schedule of their passerine prey.

Juvenile (hatching-year) birds precede adults. Fall migration for Sharp-shinneds begins during late August, peaks during the last week of September or the first days of October, and continues into early December. Spring migration occurs primarily in April or the beginning of May and falls off quickly so that it is all but completed before mid-month. Owing to their more extensive and more northerly distribution, more Sharp-shinned Hawks than Cooper's are seen at most hawk-migration junctions, although timing may shift the balance. At times, in certain places, Cooper's Hawks will outnumber Sharp-shinneds.

In autumn, in the East, Cooper's Hawk migration runs about two weeks later than Sharp-shinned migration and peaks two weeks earlier in the spring. In the West, the two species' migratory timetables are similar. Like Sharp-shinneds, juvenile Cooper's Hawks precede adults in the fall.

Adults and juveniles of both species appear to favor different migration routes. Juvenile accipiters constitute most of the birds recorded at East and West Coast hawk-watch sites, while adult birds account for more than half of the accipiters passing interior hawk-watch sites.

The reason for this disparity is not known. Possibly, adult birds, having learned about water barriers the previous fall and having pioneered a more direct return route in the spring, avoid coastal areas on their second and subsequent trips south. Young accipiters, having no experience with water barriers, reach them by accident and concentrate as geography directs. From a practical standpoint, divergent migration routes (like staggered migratory timetables) must benefit the species. While coastal routes are longer and more energetically expensive, they also provide younger and less experienced accipiters a surfeit of equally young and inexperienced prey insofar as a very high percentage of small songbirds migrating along the coast in fall are also inexperienced juveniles.

Adult birds, more adept at catching prey, can afford to migrate later and take the more direct, more energetically efficient, but less prey-saturated routes.

Numbers and ratios of adults and juveniles vary over the course of multiple years. Some years, or for some periods of years, there may be proportionately larger numbers of juvenile birds than average. These fluctuations are tied to both the health and stability of the species' breeding population and breeding productivity in any given year.

For example, when DDT depressed nesting success among raptors, the number of adults among migrating raptors remained high (for a time), but the number of juvenile birds passing hawk-watch junctions was low. When DDT was banned, nest productivity boomed and the proportions of juvenile birds in the migrant populations increased.

Northern Goshawk is an irruptive migrant. The number of birds moving south in any given year is closely linked to the abundance of prey in breeding areas. Some goshawks, most commonly juvenile males, make moderate to fairly extensive migrations every year. Adults and many, perhaps most, juveniles commonly remain in northern woodlands.

But in years when grouse and snowshoe hare numbers fall, great numbers of juvenile and adult Northern Goshawks vacate their breeding territories and retreat south. During these invasions, the birds may occur anywhere in the United States and northern Mexico. These invasions may be extensive, involving large geographic areas (for example, in 1972, a goshawk invasion occurred over most of North America). They may also be regional. An invasion that involved the western Great Lakes in 1992 did not extend to New England or the East Coast.

Autumn goshawk migration is late by accipiter standards. It begins in late August (Goshute Mountains, Nevada) and extends into mid-December (Cape May, New Jersey), though the peak migratory period in the intermountain West is from late September to mid-October. In the East, most birds are seen mid-October to early December. Like Sharp-shinneds and Cooper's, juvenile goshawks precede adults.

In spring, goshawks return early. Northbound movement begins in February and may extend into May. Most birds pass Great Lakes sites from mid-March to mid-April.

Of the three North American accipiters, Sharp-shinned is the earliest riser, sometimes beginning its day's passage before sunrise. Cooper's Hawks and Northern Goshawks start their journey somewhat later — possibly to hunt first or, more likely, to wait for thermals to perk. During migration, an early-moving (not hunting, but migrating!) accipiter is most likely a Sharp-shinned.

By midafternoon, accipiter migration tapers off. Birds devote more time to securing prey than covering ground.

IDENTIFICATION

Distinguishing one accipiter from another is challenging. Even veteran observers do not always agree. All three accipiters have roughly similar shapes, with fairly short round-tipped wings and long narrow tails. Juvenile birds of all three species have similar plumage patterns — brown above, pale and streaked below. Of the adults, only the goshawk is distinctly different. The plumage of adult Sharp-shinned and Cooper's hawks are virtually identical in the field.

The three species are separable by size, and an accurate estimate of an accipiter's basic size (big versus small) can simplify the identification process.

But efforts to distinguish solo accipiters by size alone are mostly unreliable and complicated by several factors. First, size and distance are difficult to gauge in the field. Second, size differs greatly between males and females within each species. Female accipiters are about one-third larger than their male counterparts. Finally, adults and juveniles have different proportions. Juveniles appear slightly longer winged and longer tailed than adults. The result: Juvenile accipiters appear rangier (and thus larger), while adults appear more compact (and thus smaller).

The variability, of course, contributes to the confusion of hawk watchers. The differences in size and shape *within* each species blur the distinctions *between*

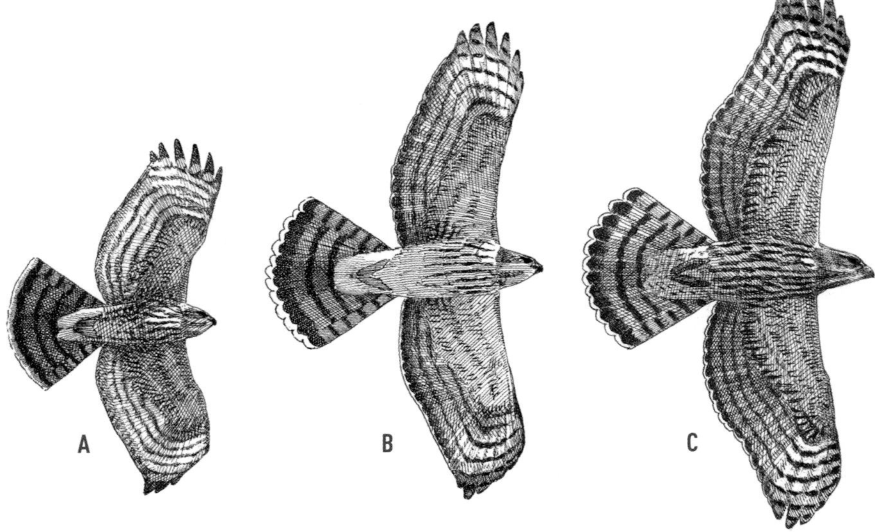

Accipiters: soaring juveniles of all species. Males are shown; females are significantly larger than males but still rarely overlap other species in size (female Sharp-shinned is nearly as large as male Cooper's; female Cooper's nearly as large as male goshawk and as large as Broad-winged; female goshawk as large as some Red-taileds). Adults of all species appear shorter tailed and longer winged than juveniles. Sharp-shinned Hawk **(A)**: short broad wings that are pushed forward in soar, all edges curved; head small; wingtips often thrust forward of head; tail with square corners. Body and underwing coverts heavily streaked; narrow gray tip on tail. Cooper's Hawk **(B)**: longer wings, head, and tail; tubular body. Wings are held straight in soar, with straighter edges and narrower tips than on Sharp-shinned; head is long; tail long and rounded. The body and underwing coverts are white with fine streaking, contrasting with the darker, orange-brown head; tail has a bold white tip. Northern Goshawk **(C)**: long wings, very broad in the secondaries and narrow at the tip; broad, heavy body but proportionally small head; broad tail. Body and underwing coverts are dingy white with heavy streaking. Flight feathers may appear darker than underwing coverts.

species. Not only do female Sharp-shinned Hawks approach male Cooper's Hawks in size, but adult Cooper's Hawks have proportions (shorter tail, shorter broader wings) that approach juvenile Sharp-shinned Hawks in shape!

Difficulties notwithstanding, identification of accipiters is no longer the impossible feat that daunted pioneering hawk watchers. A number of clues (and a few blatant field marks), taken together, make accurate identifications possible. While there are plumage characteristics that help distinguish one species from the next, accipiter identification relies heavily upon shape, relative proportions, and manner of flight. Plumage considerations, while very useful, are more supportive.

Don't rely upon just one or two field marks. In the field, none deserves to be called diagnostic or infallible. Use all the hints and clues that your study of the bird can amass, and compound them in your mind. It is not uncommon for a bird to show one or two characteristics that suggest one species and two or three that suggest another. An example might be a mostly classic Sharp-shinned Hawk with a small head, short wings, and quick, snappy, frequent wingbeats, but with a distinctly rounded tail.

Sharp-shinned tails are variable (and some are quite rounded). In this case, an

An adult Sharp-shinned Hawk from above as it darts by a hawk-watch site. This is a view most often enjoyed by watchers on ridge tops. KK

identification of Sharp-shinned wins on points. Remember, no one can be right all of the time, and no one who has watched hawks for any length of time expects to be.

Accipiter identification is challenging. It gives new hawk watchers the chance to earn their spurs and experienced hawk watchers a target for their skills. When the day comes that you can honestly say that you feel comfortable telling Sharp-shinned Hawks and Cooper's apart, you can call yourself a pretty accomplished hawk watcher.

THE GENERIC ACCIPITER

Accipiters are short-winged, broad-winged, and blunt-tip-winged raptors with long, narrow, rudderlike tails. Size varies greatly among species — from the robin-sized male Sharp-shinned to the Red-tailed Hawk–sized female Northern Goshawk. Juvenile birds are brown above and pale below, with patterned spotting/streaking (particularly on the chest). Adults are gray-brown to slate blue above and orange barred below (except in the case of goshawk, which shows fine gray vermiculation).

All accipiters, both adult and juvenile, have alternating dark and light gray bands on their tails. All juvenile accipiters show a pale stripe above the eye, and all ages show white undertail coverts. All fly using a series of flaps interspersed with a glide (more flaps and shorter glides than most buteos).

Sharp-shinned Hawk

The Sharp-shinned Hawk is a small, compact, scrappy raptor. As a breeding species, migrant, or wintering bird, it can be found during some part in the year everywhere in North America south of the tundra. A common breeder in northern and higher-elevation forests, and one that entirely vacates northern breeding areas, the Sharp-shinned is often the most common accipiter migrant at many hawk-watch sites (and, in places, the most common raptor migrant). Small bird-catching specialists that forage in forest interiors as well as brushy edges, Sharp-shinneds also haunt backyard bird-feeding stations, where they sometimes take prey up to the size of jays (though they favor birds in the chickadee and House Finch class).

In hunting mode, the Sharp-shinned maneuvers deftly through woodlands, following the contours of hedgerows and moving quickly in and out of breaks in the foliage, hoping to surprise prey. (Cooper's Hawks are much more likely to hunt, and perch, in open country such as marshes and open meadows.) Perch hunting is also an old Sharp-shinned trick (as people who feed birds in winter know well).

A Sharp-shinned's principal advantages are surprise, rapid acceleration, and reflexes so quick that the hawk's movements seem not to mimic the escape maneuvers of prey but to mirror them. Its weaponry consists of long-toed, needle-tipped feet attached to legs that are exceedingly long and fast. The term *boardinghouse reach* aptly applies to Sharp-shinneds.

If a Sharp-shinned does not secure prey after a brief acrobatic chase, however, it ends the pursuit. The long, spirited, winding chase through brush and bramble

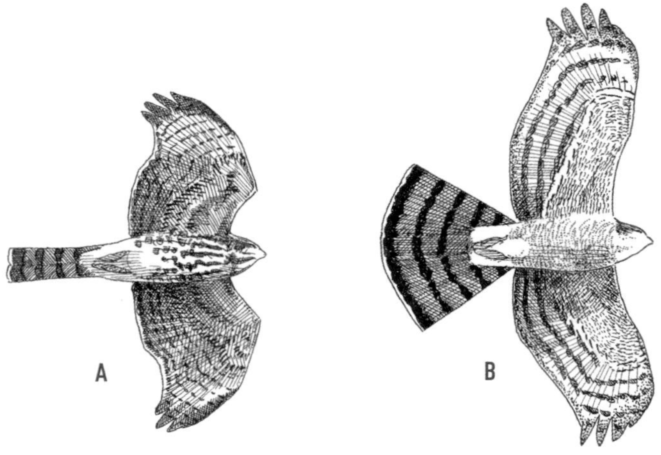

Sharp-shinned Hawk, underside. Juvenile gliding **(A)**. Adult soaring **(B)**. Wings always pushed forward, with distinct S-curve on trailing edge; tail narrow and square tipped; wings average narrower and tail slightly longer on immature. Immature heavily streaked and spotted on body and underwing coverts, appearing dirty brown and lacking the contrast between head and body evident on Cooper's. Adult plumage nearly identical to Cooper's except for usually narrower pale tip on tail.

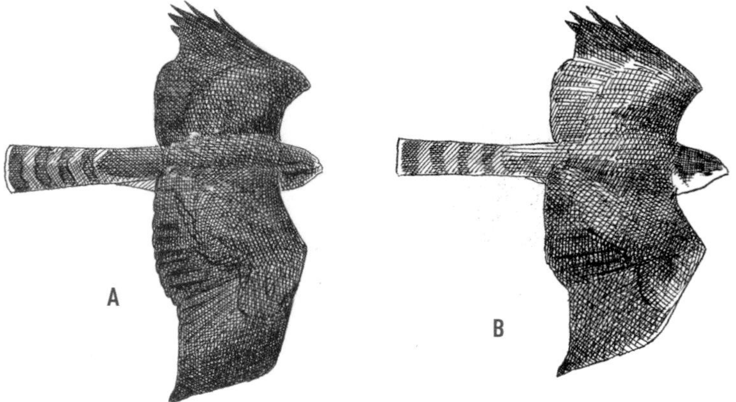

Sharp-shinned Hawk, upperside. Juvenile female **(A)**. Adult male **(B)**. Adult is blue-gray with blackish wingtips and tail bands, palest on rump. Juvenile usually appears uniformly dark and brownish above. Other juvenile accipiters have more buffy spotting on back and appear paler and more patterned. Adult plumage is extremely similar to Cooper's, but may show more extensive gray on nape. Adult male Sharp-shinneds and Cooper's are blue-gray above with darker primaries (females more uniformly grayish) so sex of adults can often be guessed by color, which makes judging size more reliable.

perpetuated by early nature writers is a myth. Sharp-shinneds may, however, sit and wait for longer periods if their intended prey takes cover in an isolated, strategically inopportune bush.

On its breeding territory, the bird is secretive. It hides its stick nests well, high in a spruce or other conifer. Caution is the hallmark of the Sharp-shinned. Although a raptor, it is not exempt from predation. Its size makes it exceptionally vulnerable to larger species of raptors, including Cooper's Hawks, Northern Goshawks, Red-tailed Hawks, and the larger falcons.

In winter, the birds vacate most of Canada and apportion themselves in forested or broken-woodland habitat from the more temperate northern states south to (at least) Central America. Increasing numbers of Sharp-shinneds seem to be gravitating to the bounty of prey species that are attracted to backyard feeding stations. Mornings and late afternoons are prime feeding times. On cold mornings, Sharp-shinneds may hunt at first light.

In migration, Sharp-shinneds are widespread, not bound to either forests or the ridges and coasts, migrating as buteos do: on a broad front. But concentrations of Sharp-shinneds do occur at these geographic leading/diversion lines and, given the right conditions, numbers can be spectacular, with Sharp-shinned numbers exceeding all other migrating birds of prey combined.

IDENTIFICATION. This raptor ranges in size between a robin and a pigeon. Males are smaller than females. Overall, the Sharp-shinned represents the compact, narrow-tailed extreme in the accipiter lineup.

The body is tubular and slight; the tail narrow and long — narrower than the body, proportionally narrower than the tail of Cooper's Hawks or goshawk.

Sharp-shinned Hawks gliding.

Wings are short, broad, and quite rounded at the tip. When soaring or gliding, narrower hands impart a distinct S-shaped curve to the trailing edge of the wing. This S-shaped configuration is most pronounced on juvenile birds. The wings of adults are more evenly tapered (recalling, though not replicating, the wings of Cooper's Hawk).

When soaring or gliding, the wrists (or elbows) jut forward, projecting almost as far as the head, which is small. Tucked in this gully formed by the jutting wrists, a Sharp-shinned's head *disappears* when viewed from a distance — it gets lost in the gully.

With its short, broad, rounded wings, long narrow tail, and indented head, the soaring and gliding Sharp-shinned appears to be all wing and tail — a flying mallet. The classic Sharp-shinned tail is square-cut across the tip and sometimes conspicuously notched. Some individuals (most notably juvenile females in the fall) have a tail that is somewhat rounded, *particularly at the corners.* But Sharp-shinned Hawk tails never appear spoon-shaped or lobed. They are, on a few individuals, simply more round than straight-cut. Spring adults are more classically straight-cut across the tails.

PLUMAGE. Adults are blue backed — blue-gray or blue with brownish highlights in females, slate blue in males. The crown is darker than the back, but not dramatically so. Underparts are orange barred — particularly on the breast. From a distance, particularly in bright light, underparts may simply appear light or pale.

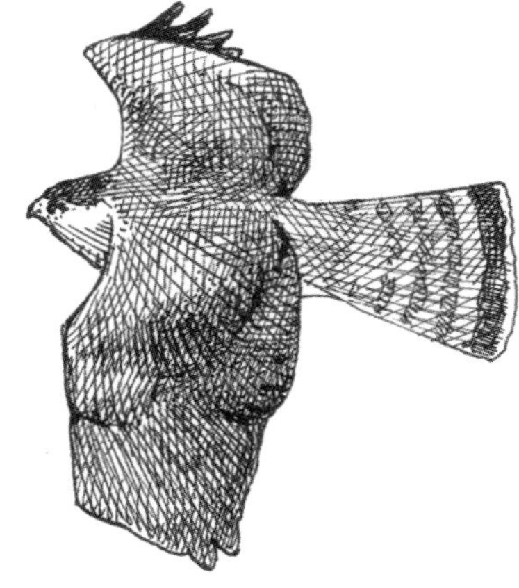
An adult male Sharp-shinned Hawk.

Juveniles are brown backed. The tone is cold (a dark brownish gray). Underparts are dirty white, overlaid with large, sloppy, brown to rufous-brown teardrops that align themselves in thick, beaded streaks on the chest and down the flanks. At a dis-

A Sharp-shinned Hawk stooping.

tance, juveniles appear dirty chested and dark or dirty flanked.

Tails on adults and juveniles are banded light and dark gray. The tip (or terminal band) is narrow, not sharply defined, and usually grayish above and off-white below (that is, less conspicuous than the broader, whiter terminal band on Cooper's Hawk).

Juvenile Sharp-shinned (left) and Cooper's hawks gliding.

IN FLIGHT. In soaring flight, a Sharp-shinned generally holds its wings straight out to the sides and thrust slightly forward. Viewed head-on, the wings show a slight droop. In a tucked glide, when the bird is riding an updraft, the wings may be angled sharply down, with the tail cocked up — something Cooper's Hawks (and American Kestrels) do not do.

The small size of the Sharp-shinned translates to a rapid wingbeat. The flaps are quick and snappy, given in a flurry too rapid to count. Each series of wingbeats is followed by a short glide. The birds are light, buoyant, and nimble in the air. In moderate winds, when in a glide, they jerk or bounce in the air!

Sharp-shinneds are high-strung and energetic. They flap frequently, almost compulsively. Even when soaring, rising in a thermal, the birds cannot refrain from bursts of flapping. Cooper's Hawks, by comparison, love to soar and are loath to flap.

In fall, Sharp-shinned Hawks frequently travel in pairs or small groups (up to eight or more at places like Cape May, New Jersey, and Kiptopeke, Virginia). They are feisty and frequently harass other raptors as large as or larger than themselves. In thermals with mixed species or places where raptors are concentrated, Sharp-shinneds commonly fly higher than species that might pose a threat.

1. An adult female Sharp-shinned Hawk soaring, showing barred orange underparts and banded tail. A fairly rangy individual, but note the S-shaped curve on the trailing edge of the wing (i.e., broad secondaries and narrow hand). Note, too, the indistinct grayish tip on the tail. Compare to Cooper's Hawk #2. Utah, October JL

2. An adult male Sharp-shinned soaring. A more richly plumaged and classically shaped individual than #1; only a few individuals have coloration this bright. Note the squared-off (and in this case notched) tail and narrow grayish tip to the tail. Utah, September JL

3. An adult female Sharp-shinned from above, riding (in fact, being buffeted by) strong winds. Grayish above, with a narrow grayish tip on the tail. The color of this individual's terminal band is the subtle and gray extreme; many birds show somewhat whiter, better-defined terminal bands. This bird's tail is slightly up-cocked, something Sharp-shinneds do in strong updrafts for stability. Gliding Cooper's Hawks have tail parallel to the body. Cape May, October KK

4. A gliding adult Sharp-shinned high overhead. This shape is classic. Note the short rounded wings with the S-shaped trailing edge. Very narrow, square-tipped tail. One thing this shot shows well is wings with wrists jutting forward, causing the hawk's small head to get lost in the gully. Step back from the book and the bird's head will merge with the wings, making the bird look like a flying wooden mallet. Nevada, September JL

5. The dorsal view of a juvenile Sharp-shinned Hawk riding an updraft. More brownish above than an adult. The shape on this one is classic, but the reddish highlights about the head and wing coverts are less typical of this species (most are duller, grayer). Also note the shape of the head and bill. Cooper's head profile looks like a bird with a regal Roman nose. (See Cooper's Hawk #3.) That description certainly doesn't apply to this Sharp-shinned. Cape May, October KK

6. A juvenile Sharp-shinned, gliding between bouts of flapping. Compared to Cooper's, note the broader wings pushed slightly forward, and shorter, square tail. Also note the heavily streaked underparts — with streaking running the length of the body (including the flanks), not concentrated mostly on the breast like a typical juvenile Cooper's Hawk. At a distance, this streaking appears smudgy on Sharp-shinned. Utah, September JL

7. A juvenile Sharp-shinned in a high banking turn. Note, first, the overall shape. Short stocky wings, with an S-shaped trailing edge. Slender body and narrow, square-tipped tail. Now note all the streaking on the underparts. In just a moment this bird will burst into a flurry of wingbeats. Sharp-shinneds are more high-strung than Cooper's Hawks; even when soaring, they seem compelled to flap. Cape May, November CS

8. A juvenile Sharp-shinned gliding. In a tucked glide, or when riding an updraft, birds seem compact overall and wings are angled down. Owing to their small size and light wing-loading (especially males), birds in this configuration commonly bounce around a great deal, even in moderate winds. Cape May, October CS

9. A juvenile Sharp-shinned all tucked up, as it navigates a buffeting headwind or strong updraft. Note the "head lost in the gully" effect and the frequently seen up-cocked tail, as well as heavy streaking below. Cape May, October CS

Cooper's Hawk

If you took a Sharp-shinned Hawk, grabbed its head and tail, and gave a steady pull, then did the same thing to the wings, you'd have a bird that bore a passing resemblance to a Cooper's Hawk — but you still wouldn't have a Cooper's Hawk. This larger, more southern cousin of the Sharp-shinned is at home in the Canadian-zone woodlands of New York State, the wooded hilltops of Ohio, deciduous forests in the Carolinas, riparian corridors in Colorado, and urban parks in California. As eastern forests have recovered and planted trees have flourished in what were once prairie communities, Cooper's Hawks have become increasingly suburbanized. Like Sharp-shinned Hawks, Cooper's Hawks have also discovered the bounty of prey that winter bird feeding attracts to backyards and have become feeder regulars, generally taking larger prey than do Sharp-shinned Hawks. Doves and pigeons are something of a mainstay in their diet; Blue Jays, American Robins, and Starlings, too. More than Sharp-shinned Hawks, Cooper's Hawks broaden their diet to include mammalian and, in the Southwest, reptilian prey. Some individual birds become chipmunk and squirrel specialists.

It came as something of a surprise when in some state breeding-bird atlases, Cooper's Hawks, which are secretive at most times of year but particularly when nesting, were found to be among the most common (in some cases second most common) nesting raptors in the state! Lack of persecution, resulting from raptor protection laws, has certainly benefited this bird. But the Cooper's Hawk's proven ability to acclimate itself to human habitat cannot be overlooked.

In many places, Cooper's Hawk numbers have exploded. Because the bird is specialized to hunt in semi-open habitats and woodland edges, suburbia and urban areas that incorporate wooded parks and riparian forest corridors have helped the species. Preconceptions relating to the paucity of Cooper's Hawks relative to Sharp-shinneds have little standing now. At hawk-watch sites across much of North America, the ratio of Sharp-shinneds to Cooper's Hawks is between three and five to one. In parts of the West, the two species occur in equal numbers.

There are times and places where probability plays a useful part in the identification of accipiters. For example, the migratory timetables of Sharp-shinned and Cooper's hawks differ (Sharp-shinneds migrate somewhat earlier in fall and later in spring). Additionally, Cooper's Hawks appear more thermal dependent than Sharp-shinneds, so in migration they are most commonly seen between midmorning and midafternoon.

Nevertheless, playing the odds is a perilous way to approach accipiter identification. Happily, given the multitude of distinguishing characteristics, it is also unwarranted.

IDENTIFICATION. Cooper's Hawk is a crow-sized raptor, larger (in the case of females, conspicuously larger) than Sharp-shinned Hawk. This is a long, lean, lanky accipiter. Its body is tube-shaped, but sturdier and more robust than that of a Sharp-shinned. The tail is long — on average, longer and proportionally wider than the tail of a Sharp-shinned. The head, too, is large, prominent, bullet-shaped, and projects well ahead of the wing. (Imagine a curious turtle extending its head from its shell.)

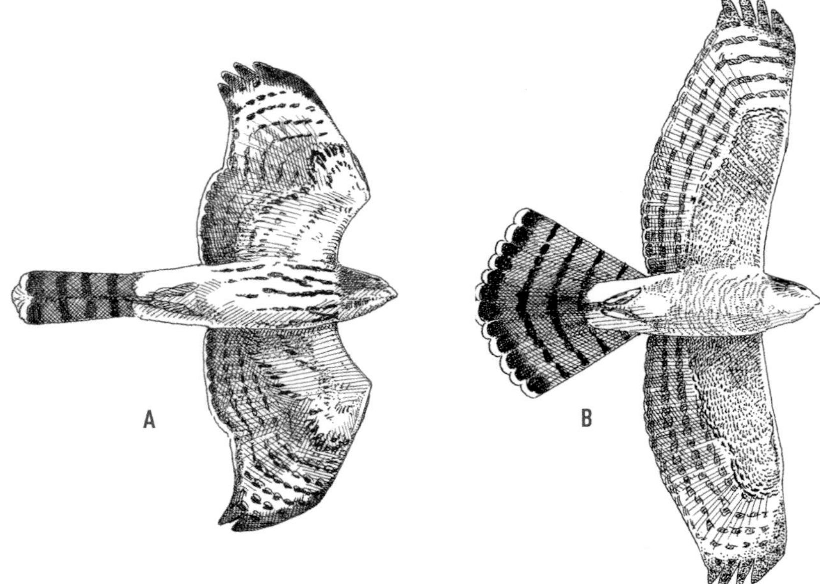

Cooper's Hawk, underside. Juvenile **(A)**. Adult **(B)**. Long tubular body; long head; wings fairly long and straight edged; tail long, club-shaped, rounded, tending toward wedge-shaped in females. Juveniles appear slightly longer tailed and thinner winged than adults. Adult has fine orange barring on body and underwing coverts. Juvenile has white body and underwing coverts, with fine dark streaking restricted to forward half of body; head solid orange-brown, contrasting with breast; underparts appear white at a distance, thereby differing from those of both other accipiters. All ages have broad white tip on tail.

The wings are long, fairly narrow, and more evenly, cleanly tapered — that is, largely devoid of the Sharpies' prominent S-shaped curve along the trailing edge. When soaring, the leading edge is straight-cut and held at a right angle to the body with little or no jut at the wrists. The overall impression is a longer-winged, cleaner-lined, more pointy-winged accipiter. This impression is even truer of adults.

If a soaring Sharp-shinned looks like a flying mallet, then a soaring Cooper's — with its projecting head; straighter, longer wings; and longer, broader tail — looks much like a flying Roman cross or crucifix. In a tucked glide, with its projecting head and long wings angled sharply back, Cooper's Hawk looks like an arrowhead. Sharp-shinneds look blunt by comparison — like an open umbrella.

Cooper's Hawk almost always shows a distinctly round-tipped tail. Sometimes the tail widens or flares out near the tip, imparting a spoon-shaped look. Sometimes, too, the tip of the tail shows a clover-shaped configuration with three distinct lobes (one in the center, one to either side). Whatever the fine points of the configuration, the tail is almost always *rounded.*

That said, some Cooper's Hawk tails are square-cut enough to be confused with a Sharp-shinned's. Second-year birds, whose white tail tip has been broomed away over the course of the winter, are particularly prone to show blunt or square-tipped tails in spring. Occasionally, Cooper's Hawk tails will have a slight cleft at

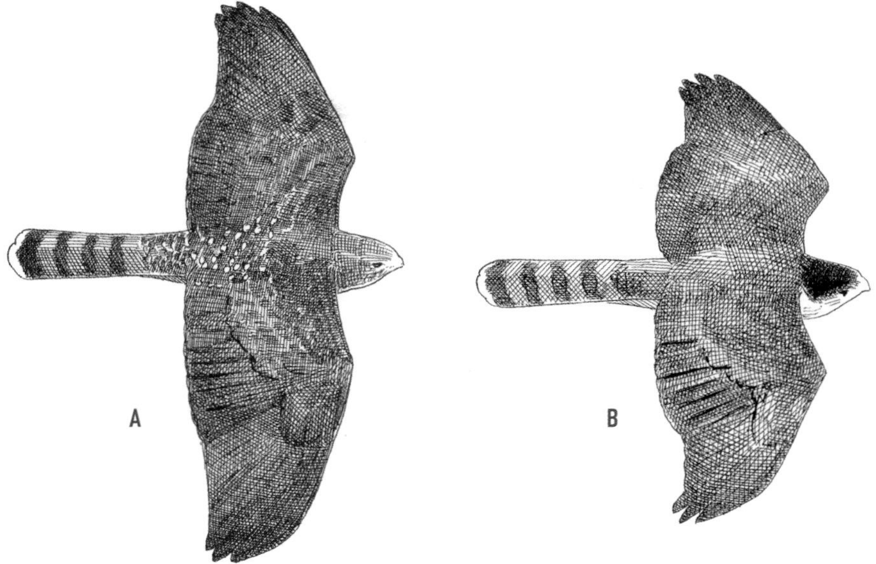

Cooper's Hawk, upperside. Juvenile female **(A)**. Adult male **(B)**. Adult must be identified by size and shape, as plumage differs from Sharp-shinned only in broader, whiter tip on tail and paler nape with darker cap, which can all be very difficult to see in flight; adult female is browner above than male, less blue, and more uniform (male has darker brown primaries). Juvenile is brown with fine buffy spotting and edging, especially on lower back and rump, appearing paler, warmer, and less uniform than Sharp-shinned; tawny head can also be obvious from above.

the tip of an otherwise very rounded tail (less acute than the V-shaped notch that adorns the tail of some Sharp-shinned Hawks).

PLUMAGE. The plumage of adults is similar to that of Sharp-shinneds — blue backed, orange breasted. Males are bluer, while females are drabber, more brownish blue. Also, and unlike Sharp-shinned, adult male Cooper's shows a dark cap that contrasts markedly with the bird's paler back, but this feature is less apparent on females.

Like Sharp-shinneds, juvenile Cooper's Hawks are brown above, cream-colored and streaked below. The upperparts are warm brown, infused with orange or tawny highlights *particularly around the nape and head.* The brown upperparts of juvenile Sharp-shinneds and Northern Goshawks show a colder, grayer cast.

The streaking on the underparts of a juvenile Cooper's Hawk is neat and fine — trickling tears more than teardrops — and restricted mostly to the chest. At a distance, the streaking disappears, giving the juvenile Cooper's Hawk a clean-chested look or a sense of pale underparts, much paler and whiter than a Sharp-shinned's. Contrasting with these white underparts is a dark head, recalling (and often called) a hangman's hood.

Accentuating the tip of the Cooper's Hawk tail is a crisply defined broad white terminal band. This band is easily seen from above and is particularly obvious from below when the bird fans its tail in a full soar. The tails of adult and juvenile Sharp-shinneds and Northern Goshawks also show a whitish tip, but it is commonly narrower, duller, and less crisply defined (and generally not visible at distances where the white-banded tail tip of Cooper's remains conspicuous).

Caution: The undertail of Cooper's Hawk is very white and prominent. It is not uncommon for observers to mistakenly project its location to the top of the base of the tail and misidentify the bird as a Northern Harrier — a bird whose size and shape are similar to Cooper's Hawk, but whose white rump patch really is on the rump.

IN FLIGHT. The wingbeat of Cooper's Hawk is slower than that of Sharp-shinned. The movement is stiff and choppy — more arthritic than elastic. All movement seems localized to the arm, not the hand. Cooper's soars in wider circles and soars more often and more extensively than Sharp-shinned Hawk. When the bird soars, the wings are sometimes held in a slight dihedral.

Heavier birds than Sharp-shinneds, Cooper's Hawks do not bounce as much in turbulent air. Instead they seem to glide, slide, or slither through the air, adding a snaky demeanor to the birds' physical qualities. Cooper's Hawk also flaps less than Sharp-shinned. When soaring, a Cooper's flaps sparingly or not at all. Soaring Sharp-shinneds seem to flap compulsively.

Cooper's Hawks tend to be solitary, unlikely to travel in pairs or small groups like their smaller cousins (although multiple birds may occupy the same kettle before continuing on). When Sharp-shinneds and Cooper's Hawks occupy the same airspace (or thermal), it is common for smaller Sharp-shinneds to fly at a higher altitude — out of harm's way.

1. A juvenile Cooper's Hawk with dinner on its mind. A "Coop in a stoop" or in a reckless tail chase are typical views for many at backyard bird-feeding stations and migratory stopover concentration points. Arizona, May NH

2. An adult male Cooper's Hawk in a full soar. The head projecting ahead of the straight-cut leading edge of the wing and the long tail give the bird the look of a flying Roman cross. Underparts are orange barred at close range but may appear just pale at great heights or distances. Utah, October JL

3. The dorsal view of an adult Cooper's Hawk. Slate gray above, with a darker cap. Note the Cooper's Hawk's Roman-nose look. On a ridge, gliding birds commonly hold their tail straight out behind their body. In gusty winds, Sharp-shinned Hawks often cock their tail up. Cape May, October KK

4. A juvenile Cooper's Hawk in a glide along a mountain ridge. Wings are well tucked but Cooper's Hawk's large head still projects. Things to note include the "hangman's hood" infused with warm or orangy tones about the cheek and nape; beaded, teardrop streaking on the breast; and bright, broad white tip to the tail. This is a particularly heavily streaked individual (seen more commonly in the West). Utah, September JL

5. A juvenile Cooper's Hawk from above. This is one long, lanky accipiter! Note the long, snaky head projecting from the straight-cut leading edge of the wing. Tail tipped with a broad, very white terminal band. Also note the warm orangy touches infused into the bird, especially the head and neck. They impart an overall warm brown cast to juvenile Cooper's Hawks. Juvenile Sharp-shinneds are typically cold brown. Utah, September JL

6. A juvenile Cooper's Hawk. Another flying cross, caught here gliding between bouts of flapping. This bird shows the "hangman's hood" and neater, finer streaking on the breast than Sharp-shinned (which makes the contrastingly dark hood stand out). Cape May, October CS

7. An adult Cooper's Hawk (left) and a juvenile Cooper's (nearing one year old) in a breeding season encounter. This is probably a territorial dispute rather than a courtship flight, as, based on size, they both appear to be the same sex. A female would appear noticeably larger than a male. Arizona, March NH

8. A soaring juvenile Cooper's Hawk. A classically shaped bird, right down to the round-tipped tail. The birds are overall lankier, more stretched out than Sharp-shinned Hawks, but adult Cooper's Hawks are typically more compact than juveniles. Cape May, October CS

9. A juvenile Cooper's Hawk in a full soar, showing long broad tail. This pale-breasted, conspicuously hooded individual is very typical (also see p. 82). Though distant, the warm or orangy tones about the head are apparent. Utah, September JL

Northern Goshawk

The Northern Goshawk is a large raptor of northern and western forests. It has the size and power and even (somewhat) the proportions of a buteo, but the hunting demeanor and killing skills of an accipiter.

The adult is striking: a gray-cloaked, charcoal-masked shadow with piercing red eyes. The bird moves through the forest in short bursts of flight. Taking a perch, it waits for prey that it then captures in an all-out aerial sprint. The bird also cruises along woodland edges and shorelines, hoping to surprise and flush prey, and is adept at using vegetation and obstacles to approach prey in a preplanned attack.

An adult Northern Goshawk.

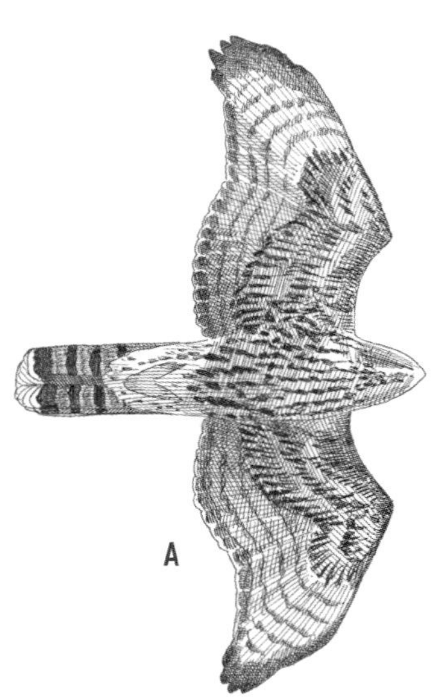

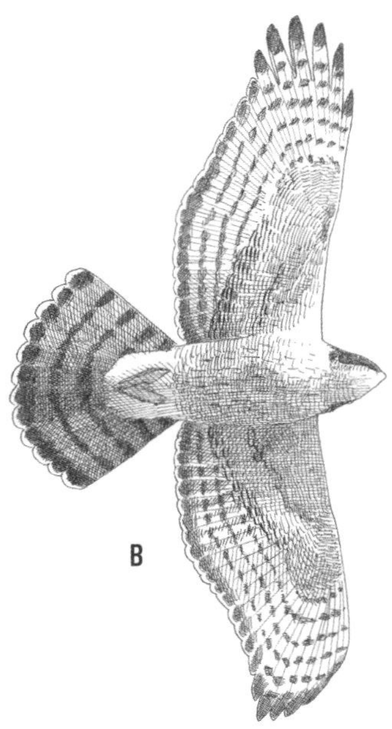

Northern Goshawk, underside. Juvenile **(A)**. Adult **(B)**. Large and heavy, with big tubular body; broad buteolike head and broad tail (shape of tail tip varies). Wings pinched at base, broad in secondaries and narrower at tip; leading edge fairly straight, but S-curve on trailing edge may recall Sharp-shinned. Juveniles slightly longer tailed and appear thinner winged than adults. Adult is easily separated from other accipiters by pale gray barring on underparts and black eye patch. Juvenile differs from Cooper's in having dingy brown body and underwing coverts with heavy dark streaking and gray-brown, not orange, head. Differentiated from Sharp-shinned by size; by dingy (not white) ground color on underparts; by lack of blotchy bars on flanks; and by white tip on tail.

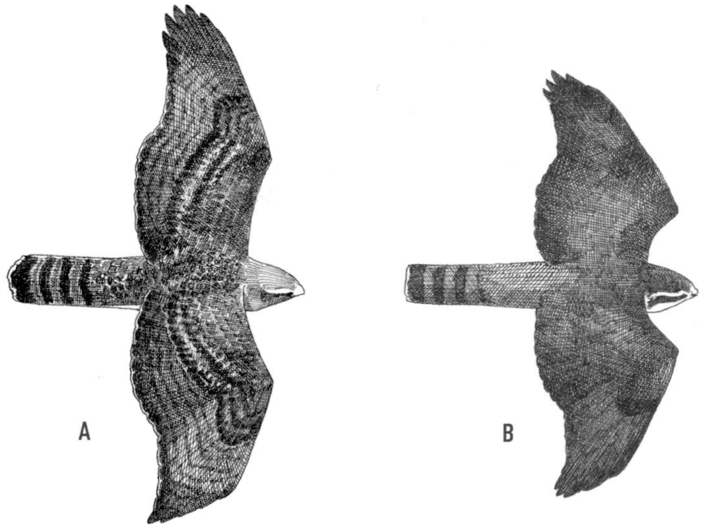

Northern Goshawk, upperside. Juvenile **(A)**. Adult **(B)**. Adult is dark blue-gray with black cap; black-and-white head pattern is unique and distinctive. Back and wing coverts on immature are distinctly spotted; all feathers edged dull buff; with prominent jagged pale bar across wing coverts. White eyebrow line may appear on other juvenile accipiters.

Unlike other accipiters, goshawks are tenacious. They chase prey, pursuing it (even on foot) as long as the opportunity for capture remains. This single-mindedness has obvious advantages for a diurnal raptor that remains in northern forests during the winter months, when prey is scarce and daylight hours are limited.

Although goshawks have historically been birds of northern and western coniferous forests, in recent years they have expanded their range into mixed and maturing deciduous and coniferous forests in the Northeast. In some places they have become somewhat suburbanized, nesting in well-forested neighborhoods and establishing themselves as the bane of Golden Retrievers and suburbanites who jog down paths in local parks and nature preserves. Maternal instincts run high in female goshawks. Whereas Sharp-shinned and Cooper's hawks are reticent and retiring, Northern Goshawks are vocal and spirited defenders of their nests.

While a few (mostly juvenile) goshawks migrate south every autumn, large numbers of juvenile and adult birds retreat south when northern prey populations crash — often much farther south than during non-irruptive years. These invasions are irregular and difficult to predict, but large irruptions are sometimes followed by a smaller echo flight the following year.

IDENTIFICATION. Northern Goshawk is a large, robust, Red-tailed-sized and buteo-shaped accipiter. From below, the body is uniformly wide along its length. From the side it seems chesty, at times pot bellied. The back is humped or hunched.

The head is broad, buteolike, wedge-shaped, configured more like Sharp-shinned than the bullet-shaped projection of a Cooper's Hawk. But the leading edge of Goshawk's wing is straight. The head doesn't get "lost" in the gully like Sharp-shinned. Nevertheless, the head of Goshawk often seems small relative to the size and proportions of the bird, imparting to the bird a "pinheaded" look.

By comparison, the tail is wide and broad. A Northern Goshawk's tail does not narrow near the base as much as the other two accipiters, the tail remains broad and uniform along the length, appearing more nearly an extension of the broad body than an appending tail — a trait that makes the bird look like "a flying stovepipe," to quote one ardent admirer.

This uniformity of breadth notwithstanding, the tail of Northern Goshawk is otherwise quite variable in length and in the shape of the tip. Juveniles tend to have a long tail with a tip that is more pointed or wedge-shaped than rounded. Adults have a shorter tail — so short (particularly in relation to the longer wings) that it may, in fact, seem more characteristic of a buteo. The tails of adults may be rounded, wedge-shaped, or blunt. Rounded or blunt are configurations shared with Cooper's and Sharp-shinned. Wedge-shaped tips are typical of young goshawks and almost unique.

Like its body and tail, the wings of Northern Goshawk are also broad, but, unlike Sharp-shinned and Cooper's, the birds appear short armed and long handed. Juveniles show the same S-shaped curvature along the trailing edge as Sharp-shinned, but the leading edge is more straight-cut.

The wings of adults are more evenly tapered along their entire length, which makes them more reminiscent of Cooper's Hawk, but the broad secondaries and long tapered hand (in combination with the robust body and short broad tail of adults and many juveniles) makes goshawk appear very buteolike — a sense that Cooper's Hawk never projects. Also, and particularly in a glide, the long, now acutely tapered hands give goshawk a falconlike appearance.

PLUMAGE. The plumage of adults is distinct and striking: gray-blue above, finely vermiculated gray below (but chest and belly project a uniform silver gray at a distance). The head pattern — dark cap, pronounced white line above the eye, dark charcoal mask — is very distinct. Females are browner tinged above and slightly more coarsely barred below. Second-year birds are mostly gray backed but have worn brown flight feathers.

The upperwing coverts are pale, forming a blue bar along the middle of the wing. The underwings are slightly two-toned, with pale wing linings that contrast with a slightly darker trailing edge (a characteristic not shared with the other accipiters).

At anything approaching a reasonable distance, an adult goshawk is easily distinguished from the blue-backed, orange-breasted adult Cooper's or Sharp-shinned. Juveniles, however, are another matter.

Like other juvenile accipiters, Northern Goshawks are *generally* brown above, variously and generously spattered with pale feathers *particularly on the upperwing coverts.* The pale coverts form a distinct pale bar on the upperwing. The lighter eye line is less distinct than on adults. The overall color tone of most juvenile goshawks is colder and grayer than the warm brown cast of a juvenile Cooper's Hawk. Some individuals, owing to a mix of blue and brown feathers, even project a purplish cast.

From below, the body and underwings are creamy to buffy (the base color buffier than that of juvenile Cooper's Hawk), overlaid with thick, often heavy, beaded brown streaks. The streaking is heavy on the breast and belly, giving juvenile birds a dirty-chested appearance. The streaking also often appears more checkerboard patterned than streaked, much more reminiscent of juvenile Sharp-shinned Hawk than the finely streaked, pale-chested juvenile Cooper's.

Finer plumage points such as a broad white or whitish eyebrow and zigzag tail pattern (giving the impression of a busier tail pattern) are useful but may be hard to see at a distance. *Note: Juvenile Sharp-shinneds and Cooper's also show whitish eyebrows, and on some juvenile goshawks this mark may be surprisingly abbreviated or indistinct.*

All in all, juvenile Northern Goshawk just looks colder and grayer above, and buffier with more (and overall darker) patterning below, than juvenile Cooper's Hawk.

IN FLIGHT. Wingbeat is heavy, deep, rowing — more fluid than the stiff, arthritic flap of Cooper's Hawk. Goshawks glide on down-drooped wings. In active flight, the wingtips draw back into very sharp points, looking very falconlike. In combination with the gray plumage, long but broad tail, and a measure of wishful thinking, a poorly seen adult goshawk might be mistaken for a Gyrfalcon.

When soaring or cruising, the flight sometimes incorporates elements of buoyant floating and stalling, giving the flight a rolling or roller-coaster sense that Cooper's Hawk does not.

In a soar, with tail fanned and long tapered wings extended, a juvenile goshawk (in particular) is more likely to be confused with a buteo — specifically, a juvenile Red-shouldered Hawk. *Rule of thumb: If you initially identify a bird as a buteo, but the closer it gets the more you realize that it is an accipiter, think goshawk.*

1. An adult Northern Goshawk as seen from a mountain hawk-watch site. Gray plumage above, prominent white supercilium (or dark eye patch) aside, note the big, broad, buteolike proportions of the bird, especially the broad but narrow-handed wings. Nevada, October JL

2. A juvenile Northern Goshawk riding an updraft. Overall more heavily and uniformly spotted below than juvenile Cooper's Hawk, with a busier, more banded tail. Utah, September JL

3. An adult Northern Goshawk gliding overhead, showing wedge-shaped head (that often seems too small for the bird), tube-shaped body and tail (the flying stovepipe), and long, broad wings that taper at the hands. This particular bird shows prominent white undertail coverts. Don't count on it! The wings of many adults appear two-toned, showing paler underwing linings and a darker trailing edge. Alaska, April JL

4. A juvenile Northern Goshawk gliding. This is a heavily marked individual. Note the pointy, wedge-shaped tail tip, which seems typical of many juveniles. No juvenile Cooper's Hawk would ever be so densely patterned. Note, too, the relatively small headed look that is more like that of a Sharp-shinned Hawk. Utah, October JL

5. A juvenile Northern Goshawk in a steep glide. Note robust, heavy body, and impression of two-toned wings (tawny coverts contrasting with grayish flight feathers). Michigan, April JL

6. A juvenile Northern Goshawk, dorsal view, showing overall brown upperparts and pale carpal bar. Note, too, the fairly obvious pale supercilium and the S-shaped configuration of the trailing edge of the wing — the product of broad arms and small, or narrow, hands. Cape May, November TJ

7. A juvenile Northern Goshawk soaring high overhead. A flying stovepipe set on bulging, broad wings that taper acutely at the hand. Sharp-shinneds have narrow tails. Cooper's Hawks have evenly tapered wings. And it's just too dark and robust to be either. It is individuals like this bird that make you wonder, "Some kind of buteo?" Cape May, October CS

8. A juvenile Northern Goshawk in a soar. The wide wings and tail make the head appear diminutive. Alaska, April JL

9. An adult Northern Goshawk in a glide. Note the large, robust tubular body and pointed wing appearance. Also note the pale plumage with fine vermiculation below. The eye patch is prominent. Nevada, October JL

PUTTING IT TOGETHER: Telling Accipiters Apart

When you see an accipiter, with or without binoculars, the first thing to try to gauge is *rough size*. Does the bird appear big or small?

If the bird appears small right at the get-go, chances are it truly is small — a Sharp-shinned Hawk. Resist the temptation to make it larger than it is.

But if the bird looks big, you need to decide how big. As big as a female Sharp-shinned or a male Cooper's Hawk? Or really big, like a female Cooper's or male Northern Goshawk? *Caution: On days with overcast skies, all birds appear larger than they really are. Calibrate accordingly.*

It is not particularly difficult to distinguish Sharp-shinned Hawk from Northern Goshawk because the two species differ greatly in size and manner of flight. This is very fortunate because at a distance, and with juveniles in particular, high-soaring Sharp-shinneds and high-soaring goshawks can appear very similar. Both have S-shaped curves to the trailing edge of their wings, both have a small-headed appearance, and both juveniles are dirty chested.

A high-flying Northern Goshawk can easily be passed off as "just another Sharp-shinned" until it joins a soaring Red-tailed Hawk (and the birds are nearly the same size) or until the bird flaps. There is a world of difference between the heavy, rowing wingbeat of a goshawk and the quick, snappy sputter of a Sharp-shinned Hawk.

Sharp-Shinned Versus Cooper's

For adults and juveniles, many observers look first at two key characteristics to distinguish Sharp-shinned Hawks from Cooper's Hawks: the size of the head, and the shape of the tip of the tail. Both marks are useful, but neither is infallible. No field mark is. *Never rely on just one or two features when making an identification in the hawk-watching arena. Use a number of clues and your chances of correctly identifying the bird increase.* Of these two standard field marks, tail shape is the less reliable.

If the bird is a Cooper's Hawk, the head will be bulky and protrude well ahead of the wing, giving the bird a bull-headed appearance. In a soar, the wings will be straight-cut along the leading edge, accentuating the prominence of the head. On Sharp-shinneds, with their smaller head, and wings that jut forward along the leading edge, the head gets "lost in the gully" (it's visually smothered by the body and wings), making high-flying birds look all wing and tail, like headless flying mallets. Cooper's Hawks — long in the wings, long in the head and tail — look like flying crucifixes.

Juvenile Cooper's (right) soaring with two Sharp-shinneds. Straighter wings, longer tail and head, and whiter flanks of Cooper's are all evident.

A behavioral observation, related to head size, also helps distinguish the two species. Often, a raptor passing a group of hawk watchers will have its curiosity aroused. If the passing accipiter must drop a shoulder or bank its body to get a look at the group, the bird is a Sharp-shinned. If the bird simply swivels its head, like a turtle looking back over its shell, and the body remains firm, the bird is a Cooper's Hawk (or a goshawk).

If the bird's tail is deeply notched or square, the species is likely a Sharp-shinned Hawk. If the tail is somewhat rounded, it *might* be a Sharp-shinned Hawk. In juvenile birds, about one in every ten Sharp-shinned Hawks has rounded corners to the tails. About one in every hundred Sharp-shinned Hawks has a very rounded tail.

Most Cooper's Hawks have *very* rounded tails, but you will occasionally find a Cooper's Hawk with a square-tipped tail. It is so infrequent that it generally rates a mention at a hawk watch—an exclamation similar to, "Huh, there's a square-tailed Coop."

While you are studying the tip of the tail, notice the light terminal band. If the band is narrow, grayish or dirty white, and not sharply defined, the bird is very probably a Sharp-shinned Hawk. If the terminal band is broad, very white, and very sharply defined, or if the terminal band remains conspicuous on very distant birds, the bird is almost certainly a Cooper's Hawk.

In terms of tail length, and allowing for normal variation, Cooper's Hawks *appear* to have longer tails than Sharp-shinned Hawks, *particularly* in juvenile birds. Whether this is indeed the case or a visual illusion caused by the different cut of the trailing edge of the wing or differences relating to the relative proportions of wing to tail doesn't matter.

Hawk identification isn't concerned with how things *are*. It is concerned with how things *appear*.

In terms of overall gestalt—the flash impression that veteran hawk watchers gain from hours spent studying head shape, wing length, tail length, and how these features integrate to form a composite whole—Cooper's Hawks appear lanky. Sharp-shinneds are more compact.

Caution: Adult Cooper's Hawks are more compact than immature birds. In overall shape, the difference between Sharp-shinned and Cooper's is diminished. While adult female Cooper's are much larger than Sharp-shinneds, adult males are not, and identification may, therefore, be confusing.

While adult male Cooper's Hawks may have a short tail, the tail will seem disproportionately broad—broader than the narrow, tongue-depressor-shaped tail of Sharp-shinneds.

One or two (or three) other traits are useful for separating Cooper's and Sharp-shinned hawks. The thick blotchy streaking on the underparts of most juvenile Sharp-shinned Hawks makes them look dirty chested. The neater, finer streaking found on juvenile Cooper's Hawks disappears at a distance, making the underparts appear clean and white.

Note that the dirty-chested versus clean-chested distinction applies only to juveniles. Adult Sharp-shinned and Cooper's hawks are orange barred below and, given distance and bright sunlight, both birds will show light underparts.

The pale underparts of a juvenile Cooper's Hawk contrast with and accentuate the bird's dark brown head. As a result, juvenile Cooper's Hawks commonly appear hooded. Also, the streaking on Sharp-shinneds extends down the sides, often coalescing into bars on the flanks. Juvenile Sharp-shinneds therefore look dirty flanked.

The head of an juvenile Cooper's Hawk also has an orange cast, and this warm tone is often reflected in the upperparts as well. Sharp-shinneds are colder toned, more cold brown than warm brown at a distance.

Wing-on, when a Sharp-shinned rides down a ridge in a sharply tucked glide, the tail is cocked up at an angle. A Cooper's Hawk riding an updraft holds its tail horizontally. (American Kestrels hold their tails horizontally as well.) Kestrels and Sharp-shinneds are about the same size and when riding an updraft, and seen wing-on, may sometimes be confused.)

For those whose focus is birds that fly at distances where the usefulness of shape and plumage are diminished, Sharp-shinned and Cooper's hawks may be distinguished by their manner of flight. If the bird has a quick, snappy wingbeat (at a distance it may recall a flickering candle flame), the bird is a Sharp-shinned. If the wingbeat is slower, stiffer, and choppier (with the motion centered at the arm and shoulder), the bird is a Cooper's. Because Sharp-shinned is smaller and lighter than Cooper's, it gets kicked around by the wind. Cooper's Hawks are steadier, more even keeled.

In a strong updraft, Sharp-shinneds bounce. Cooper's Hawks slide, slither, or glide. And while both birds make ready use of thermals, Cooper's Hawks seem addicted to them (and so will soar more readily and flap much less than Sharp-shinneds) and, on days when Sharp-shinneds are pinned to the ridge, Cooper's Hawks will be high and wide of the ridge.

Finally, Sharp-shinneds are also somewhat gregarious and will sometimes show up in pairs, trios, or small groups. Cooper's Hawks are solitary (although two or more may share, for convenience's sake, an opportune thermal, the birds do not travel together). A single accipiter might be either a Sharp-shinned or a Cooper's. Two or more accipiters flying (not soaring) amicably, in closer proximity, are almost certainly Sharp-shinneds.

Hint: If your identification gets all wrapped around the axle (meaning you hem and haw yourself into a corner), try rebooting the identification process. Look away from the bird. Come back at it fresh. Very often, with a fresh first glance, the bird will show something or express itself in a way that makes it eminently a Sharp-shinned or a Cooper's Hawk (a sense or hint that you would have second-guessed into inconsequence a moment before).

Cooper's Versus Goshawk

Sharp-shinned Hawks and Cooper's Hawks do not overlap in size, nor do Cooper's Hawks and Northern Goshawks — in overall length, wing length, or weight. The largest female Cooper's Hawk will always be smaller than the smallest male goshawk. Nevertheless, distance diminishes dissimilarity since a difference of two or three inches isn't conspicuous at a quarter mile. And if size alone were sufficient for field-identification purposes, there would be no grounds for confusion.

A quick two-step process can often differentiate Cooper's Hawk from Northern Goshawk:

Step 1: *Is the bird an adult or a juvenile?* An adult goshawk, with its pale blue-gray upperparts and light gray underparts — and also with its distinct facial pat-

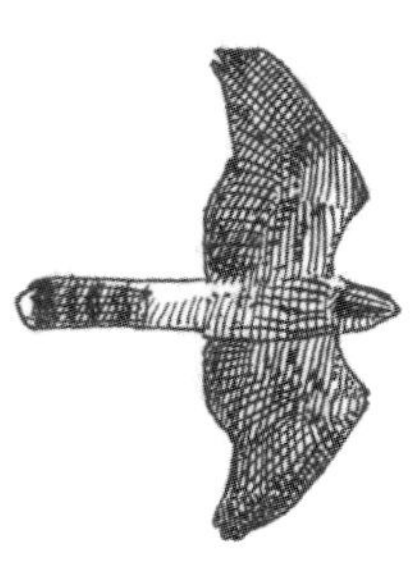
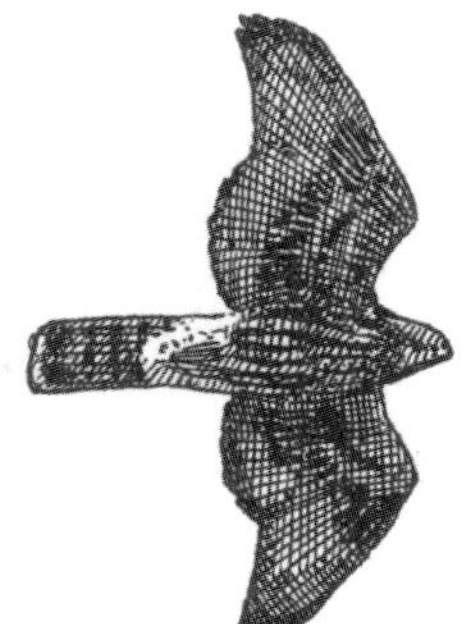

Juvenile Cooper's Hawk (left) and Northern Goshawk (right) gliding. Stockier and broader proportions and "dirty" color of goshawk are apparent.

tern — is about as different from the slate blue, orange-breasted adult Cooper's Hawk as nature can contrive. If the bird is a juvenile (if it is brown backed with streaked underparts), however, go on to . . .

Step 2: *Are the underparts heavily streaked (dirty), or are the underparts faintly streaked (clean and white)?* If the chest appears dirty at a distance, think goshawk. If the underparts are pale and light, it's likely Cooper's.

After you've made your initial identification, here are some reinforcing characteristics to note.

The head on a Cooper's Hawk is heavy and long, suggesting the head of a curious turtle. The head on a goshawk is proportionately smaller, less projecting, and wedge-shaped. At times, if the bird is stretching its neck in order to study something, the effect is still wedge-shaped, but it's a narrow wedge.

The tail of a goshawk is broad — buteolike. It seems less like a tail than an extension of the very broad body. The tail of a Cooper's Hawk really looks like a tail — narrower than the body, especially just behind the wings.

The wings of a goshawk are short and wide at the arm and narrow and long at the hand, with elements of the S-shaped curve on the trailing edge seen on the wing of Sharp-shinned. The very wide secondaries give the wing a buteo-esque appearance. The severely tapered hand distinguishes it as a goshawk. Cooper's Hawk, by comparison, is shorter handed and has a narrower, trimmer, more evenly tapered wing.

The overall gestalt of Cooper's Hawk is long, lean, and lanky; goshawk is heavy, bulky, broadly proportioned, and *buteolike.* The wingbeat of goshawk is deep, heavy, powerful, and rowing, but surprisingly quick for such a large bird. The wingbeat of Cooper's Hawk is stiff, arthritic, and more shallow.

Several plumage characteristics of juvenile birds are worth noting. Juvenile goshawks are overall paler brown, though the upperparts may have a gray or purple sheen. Juvenile Cooper's Hawks are warm toned, particularly about the head. Juvenile goshawks have a prominent pale (white or tawny) bar running along the upperwing, whereas Cooper's Hawks do not.

Finally, and as noted at the beginning of the discussion, the chest (and most of the lower body) of juvenile goshawks is heavily streaked, sometimes checkerboard patterned. Juvenile Cooper's Hawks are finely barred and appear clean chested or white below at a distance.

Falcons

Birds That Measure Distance by the Horizon

SPECIES
American Kestrel, *Falco sparverius*
Merlin, *F. columbarius*
Peregrine Falcon, *F. peregrinus*
Prairie Falcon, *F. mexicanus*
Gyrfalcon, *F. rusticolus*

An introduction to any bird bearing the name *falcon* seems superfluous. Even a person who has never seen one knows that a falcon is a very fast hawk. These are blade-winged sentinels standing post in places where distance is measured by the horizon. Falcons are at home on the Arctic tundra, prairie sage, desert and tidal flats, plowed field, open marsh, barrier beach, and, during migration, offshore waters—anywhere that the shortest distance between two points is a straight, unencumbered line.

Falcons are, for the most part, bird-catching specialists. Unlike accipiters, which are bushwhackers, falcons are adept at capturing prey in wide-open spaces where prey can run, but not hide. In the natural arena, predator versus prey, the element of surprise, coupled with the falcon's superior speed, is generally enough to tip the balance in the falcon's favor. If prey is not secured in the first attempt, the superior agility of many smaller birds is balanced or nullified by an absence of cover and falcon tenacity.

That said, not all prey taken by falcons is airborne or, for that matter, even capable of flight. The American Kestrel is a champion mouser, and Prairie Falcons thrive on ground squirrels. But falcons, for the most part, secure avian prey, and they do so in the air. Few of Earth's creatures are so accomplished in so difficult a contest.

Five of North America's six species of falcons are migratory. Kestrel, Merlin, Peregrine, and Gyrfalcon migrate (or wander) great distances. The western Prairie

LEFT: A juvenile Merlin. This bird is banking, with a partially spread tail. Compare this classic long-winged falcon shape to the less rangy shape of accipiters. KK

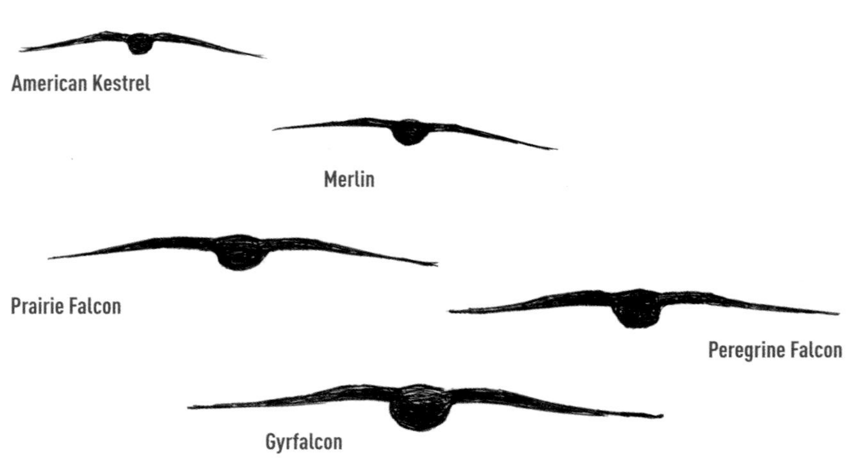

Five falcons in typical head-on flight profile. All hold their wings nearly flat, with strong shoulders and very pointed tips without upswept "fingers."

Falcon is a partial migrant, withdrawing from high altitudes and the Canadian portions of its breeding range and relocating across the western prairies and south, even into portions of Mexico. The Aplomado Falcon, a southern species recently reintroduced, is mostly sedentary and treated elsewhere in this book.

MIGRATION

The falcon's direct approach to life carries over into migration. The mountain ridges and their associated updrafts concentrate buteos, eagles, accipiters, and vultures during the fall but fail to attract falcons in the numbers that appear at coastal hawkwatch sites. Falcons are designed for sustained long-distance travel, using their own powers of flight (augmented by thermal lift). They have no real need for ridge updrafts and so are less inclined to use them. Spring observations from interior ridge sites disclose that most falcons cross ridges that lie astride their path even when the course of these ridges varies from their path by small degrees. (In fall, falcons seem a bit more tolerant of the vagaries associated with mountain-ridge travel and will use ridges more.)

For the most part, falcon migration is conducted over a broad front, spring and fall. Migrating birds are widely dispersed and not easily seen until the birds are deflected by, and concentrated along, coastal areas.

By far the greatest concentrations of migrating falcons occur down the Atlantic and Gulf coasts in the fall. While many falcons make their passage offshore, great numbers may be seen along barrier islands and, particularly, at peninsula concentration points such as Cape May, New Jersey; Cape Charles, Virginia; and the Florida Keys. Other autumn concentration points include Great Lakes sites such as Hawk Cliff, Ontario; and Duluth, Minnesota. On the Pacific Coast, the Golden Gate Raptor Observatory, near San Francisco, offers a good place from which to observe falcons and other raptor migrants.

In spring, with populations tempered by winter attrition, respectable numbers of falcons may be seen along the Texas Coast; along the Atlantic Coast at Sandy Hook, New Jersey, and Fort Smallwood, Maryland; and along Chesapeake Bay. Good numbers also occur at Great Lakes sites like Whitefish Point, Michigan; Braddock Bay, New York; and Derby Hill, New York.

The question is, if falcons are not particularly hydrophobic, why are their migrations so coastal in nature? First, while most falcons are not daunted by flying over water, this doesn't mean that they are aquatic. Even falcons must rest and feed during migration and, with the exception of some (perhaps most) adult male Peregrines, which appear to make nonstop water crossings up to several thousand miles, most migrating falcons need someplace to rest during migration, and large bodies of water are short on perches.

Also, coastal areas offer two advantages to falcons — one relating to habitat, the other to prey. Coastal areas, characterized by beaches, marshes, and open water, are ideally suited for hunting falcons — far more so than forested interior regions. In addition, in autumn and spring, coastal areas offer a surfeit of prey, especially migrating passerines in the fall (many of which are inexperienced hatching-year birds) and shorebirds (mainstay prey for several falcon species).

Another aspect of falcon migration that remains somewhat speculative relates to nocturnal migration. Through radiotelemetry, it has been established that some Peregrines do migrate at night. At Cape May, both Peregrines and Merlins are commonly seen coming ashore at *first light,* and on multiple occasions, numbers of American Kestrels have been seen flying at altitudes beyond the reach of the unaided eye before sunrise (when no kestrels were seen to be migrating at lower altitudes).

To what extent, and under what conditions, falcons migrate at night is unknown — and quite beyond the scope of this book. Identifying falcons is challenging enough when they are visible, much less cloaked in darkness.

Fall falcon migration is protracted. Kestrels, principally adults in heavy molt, begin reaching coastal concentration points in mid-July. Adult kestrels may continue to migrate into December. While some Merlins and Peregrines are migrating by late August, for most falcons, the peak of migration falls between mid-September and mid- to late October. Prairie Falcons migrate somewhat earlier than this (late August through mid-October); Gyrfalcons, well south of their Arctic breeding territories, are more likely to be encountered from mid-October into December, but some Gyrfalcons appear to be seminomadic (or have a loose and exaggerated sense of territory) and may wander many hundreds of miles during the winter months.

In spring, over much of North America, falcon migration is compressed into two months — from mid-March to mid-May.

IDENTIFICATION

Unlike accipiters, whose plumages differ in degrees and shades, the falcons have plumages that easily distinguish them from one another and from other raptor species. Given good light, close proximity, a good angle, and enough time, perched and flying birds surrender their identities without a struggle. Only occasionally do the old textbook field marks fall short — when, for instance, the occasional large female Merlin sporting a pronounced mustache is transformed into a Peregrine, or

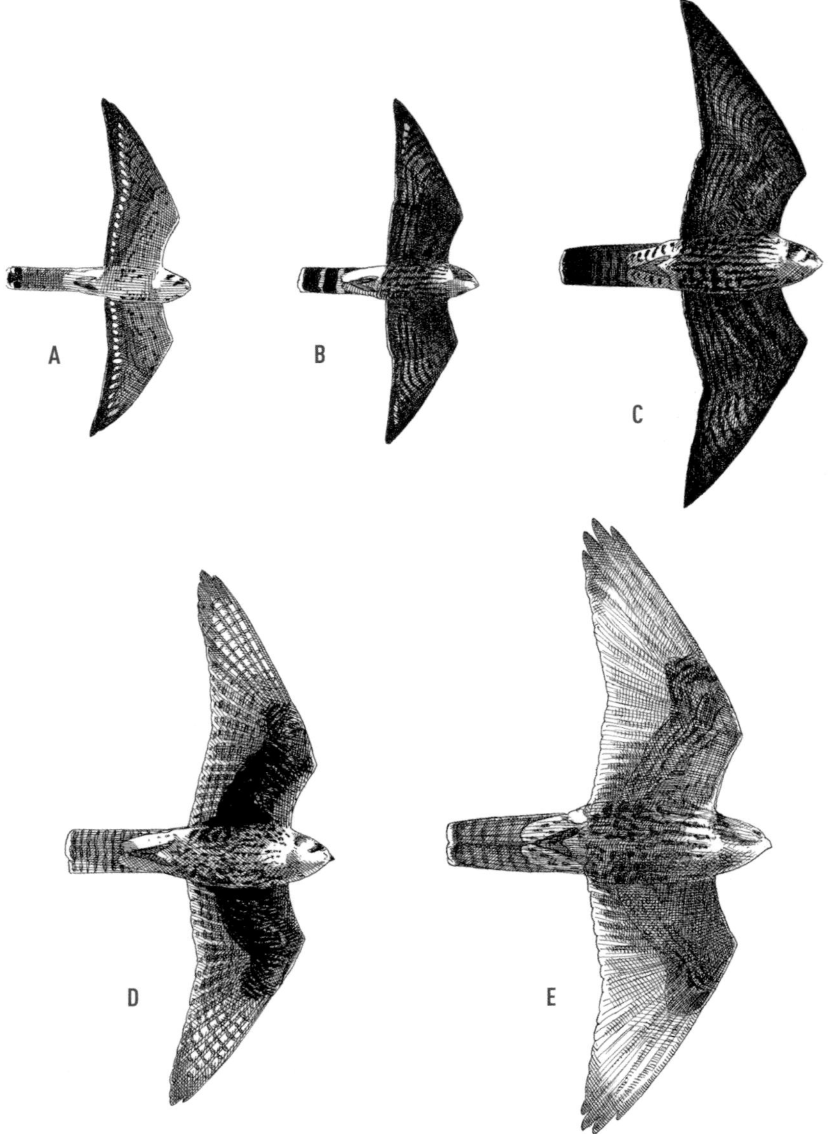

Falcons: gliding juveniles of all species, backlit. American Kestrel **(A)**: long narrow wings curved back in sickle shape; slight body, narrow tail, small head; overall pale; male shows line of translucent dots on trailing edge of wing. Merlin **(B)**: only slightly larger than Kestrel, with broader, straighter, more angular wings; shorter, broader tail; heavier body and larger head; overall dark with white throat and buffy undertail coverts. Peregrine Falcon **(C)**: long, angular, sharply pointed wings; rather heavy body; tail relatively shorter than on Merlin; shape and color variable, generally dark with streaked body, pale chest. Prairie Falcon **(D)**: long slim wings, not as angular or as pointed as on Peregrine; body slimmer on average and tail longer than on Peregrine; overall pale brown with translucent flight feathers and white chest, black underwing coverts. Gyrfalcon **(E)**: broad, blunt-tipped wings; tail long and broad; head large and body very broad and heavy; plumage extremely variable, from nearly pure white to dark gray; flight feathers translucent and paler than underwing coverts; markings always muted compared with those on other falcons.

An adult Peregrine Falcon at full bore. Falcons use powered flight to a greater degree than other groups of hawks. TJ

when a particularly dark-plumaged juvenile female Peregrine is transmuted into a Gyrfalcon.

But these textbook field marks — the double mustache on a kestrel, the narrow white bands on the tail of a Merlin, the football-helmet hood on a Peregrine — quickly show their limitations in the hawk-watching arena. Plumage still plays an important role in falcon identification at the hawk watch, but on a grander scale — in terms of light areas versus dark areas, and in terms of overall color.

Even the most distinctly plumaged birds wrap themselves in anonymity when they are distant or seen fleetingly. But falcons confer a special gift to hawk watchers. Maybe it is because living in the open makes a bird honest. Maybe a bird whose life is wedded to the horizon doesn't want to entirely surrender its identity to that horizon. For whatever reason, and more so than the other raptor groups, falcon identification is closely bound to motion — the rhythm and cadence of the bird's flight.

Each species projects a number of clues that help observers pin names to distant forms. But to discerning eyes, the way a falcon signs its name is how it flies.

THE GENERIC FALCON

Falcons vary greatly in size but all have long tapered wings that are acutely pointed at the tip. Color patterns differ, but in general, falcons are dark above and barred or streaked below. The throat and upper chest are clean or less heavily marked; the face is scored by a mustache or sideburn slash that can be bold, thin, or shadowy.

Flight is direct and fast; wingbeats steady, even, and continuous. Falcons can and do use thermals for lift, and they soar very well. When soaring, they may look deceptively *un*like falcons. In general, falcons glide less than accipiters and buteos, but glides are often protracted.

American Kestrel

This Killdeer-sized (and shaped) bird of prey may be North America's best-known raptor. It is familiar to every farm youngster as the "sparrow hawk" that nests in the old sycamore down by the creek or under the eaves where the boards are warped. A bird of open country, during migration it is a frequenter of vacant lots, waterfront parks, sports complexes, and power-line cuts, where it perch-hunts from the wires suspended between towers. In winter, millions of commuters pass kestrels twice each day as the birds hover-hunt the median strips of interstate highways.

The species is sexually dimorphic: males and females have different plumages, an uncommon trait among raptors. Moreover, the birds are dimorphic from the time they fledge. Most birds of prey have a juvenile plumage that differs markedly from adult plumage, whereas plumage differences between adult and juvenile American Kestrels are slight.

American Kestrels consume a variety of prey, ranging from crickets to carrion. Insects, small reptiles, and amphibians constitute a large portion of the bird's prey in summer. In winter, kestrels shift their focus to small rodents and sparrow-sized passerines. Most hunting is done from a perch. In the absence of perches, birds frequently hover-hunt over promising cover. Most prey is taken on the ground by either a long direct approach or an aerial pounce from a hovering vantage point. But kestrels will also take birds (and insects) in the air in true falcon fashion.

At some point in the calendar year, the American Kestrel is found in all parts of North America south of the tree line (although in the Northeast it has suffered a serious and prolonged population decline). After the breeding season, it vacates nesting territories in Canada, New England, and northern plains states. Wintering birds reach Central America. Wherever it chooses to homestead, the bird requires open or semiopen country, but not necessarily "natural" land. A sizable breeding population has established itself in New York City, for example.

IDENTIFICATION. This is a slim, diminutive falcon — in both size and shape it is somewhat parakeet-like. Females are larger than males, but this is evident only in direct comparison. Plumages of males and females are different. Males are beautiful and boldly patterned — blue winged, rusty backed, the rufous-colored tail trimmed by a broad black subterminal band. Females have dark-barred, reddish brown upperparts. The underparts of both sexes are streaked and spotted (particularly along the sides). Both sexes boast two vertical black slashes on the side of the head — a mustache and a sideburn. Except for some streaking on the breasts of juvenile males (lacking in adults), the plumage of adults and juveniles is essentially similar.

Seen well, perched or in flight, the double-slash head pattern easily distinguishes American Kestrel from all other falcons.

The wings are long and narrow, the body slight and slim. The tail is *very* slim and slightly narrower close to the body. Overall, kestrels appear daintier and more fragile than other falcons. In level flight, wings are curved back in a sickle shape, not bent or angled.

In a soar, when the long tapered wings are fully extended at a right angle to the

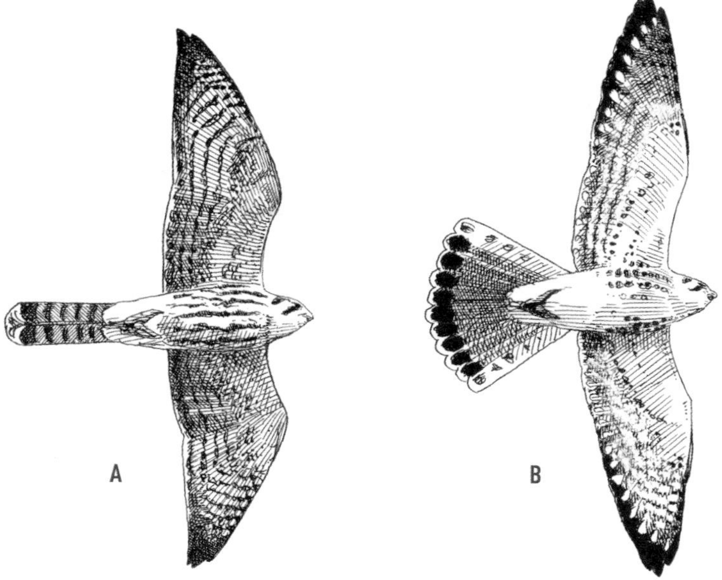

American Kestrel, underside. Female **(A)**. Male **(B)**. Long wings, narrow at base, blunt tipped in soar; long primaries curve back in sickle shape during glide. Small and lightweight, with thin body, small head, long tail. Juvenile similar to adult; sexes quite different. Body and underwing coverts appear pale; male has spots on sides of underparts, a boldly patterned rufous tail, and a row of translucent dots on trailing edge of wing. Female has streaked body, fine bars on rufous tail; only faint translucent dots on wings.

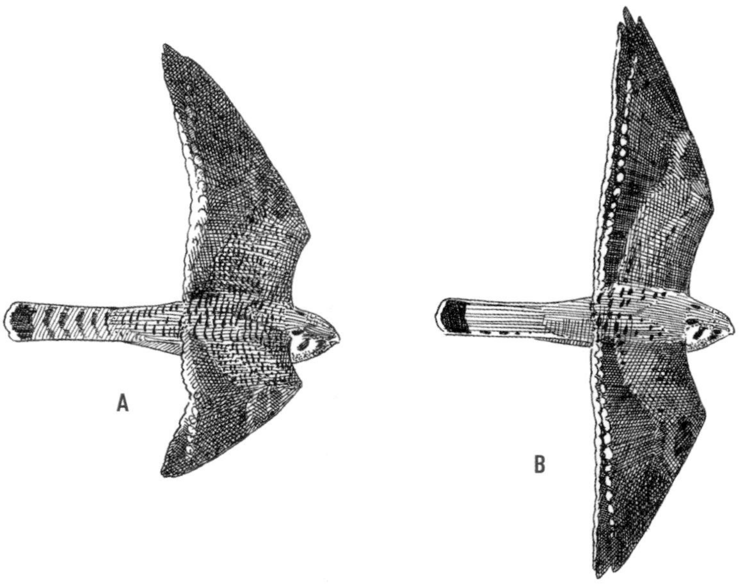

American Kestrel, upperside. Female **(A)**. Male **(B)**. Male has rufous back and tail, contrasting with blue-gray wings. Female is duller, uniformly rusty with dark barring.

body, the wings lack any hard lines or edges. The contours of the wing are rounded or curved, the angles muted or softened, making the tip appear broomed or blunted. In a glide, and sometimes when soaring, the wings appear drooped at the tip or overall slightly down-bowed. (Courting birds angle their wings sharply down.) When you think kestrel, think soft angles.

Head-on, the forehead is light. When a soaring male is overhead, his fanned red tail (set off with a dark band on the tip) is readily apparent. The underwings shine silver. On males (and on backlit females), the trailing edge of the wing is outlined in white spots (buff-colored in females). The translucent dots recall a string of pearls.

Male or female, seen from above or below, kestrels appear overall pale at a distance — lighter toned than most falcons (with the notable exception of the adult male Merlin and Prairie Falcon).

IN FLIGHT. The flight of American Kestrel is very much in accord with the bird's diminutive size and delicate contours. The flight is light and buoyant. Wingbeats are fluttery, almost dainty. The birds wander in flight — climbing and falling, wandering left, then gliding right, tacking back and forth like a sailboat in the wind. The bird's small size and light weight cause it to be buffeted by even moderate winds.

Kestrels glide more frequently than other falcons. When they glide, they float, and their forward momentum slows. This flap-and-glide pattern imparts an accipiter-like quality to the kestrel's flight (a characteristic that other falcons do not project).

Especially early during fall migration, kestrels frequently migrate in loose or strung-out groups of three to ten birds. Other falcons are solitary, paired at most.

Kestrel migration is heaviest in the afternoon, as it is for most falcons. Often, late in the day, kestrels will halt their migration and hunt. If you are studying a small, distant, low-flying falcon, and it suddenly pulls up and hovers, *stop.* There is no need to look any further for identification. The bird has just shouted, as far as discerning eyes can read, "Hi, I'm an American Kestrel."

1. A male American Kestrel eating on the wing. This species is adept at grabbing prey (usually insects) in flight. Be aware that Merlins, and rarely Peregrines, also catch insects on the wing. Cape May, October KK

2. The ventral view of a gliding female American Kestrel. The bold face pattern stands out at this distance, but note, too, the slender body and long narrow tail as well as the soft, not acute, angles along the contours of the wing. Cape May, October KK

3. The dorsal view of a male American Kestrel. Note the classic field marks: dark mustache and sideburn; reddish tail with broad dark terminal band; string of white pearls along the trailing edge of the wing. Utah, September JL

4. The dorsal view of a gliding female American Kestrel. Note the mustache and sideburn face pattern and barred reddish tail, contrasting slightly with the brownish upperparts. Note the down-bowed wings. Nevada, September JL

5. The ventral view of a soaring male American Kestrel. The striking and colorful pattern, coupled with the small size (which will be apparent in the field), render this bird unmistakable. Cape May, October KK

6. A male American Kestrel in a typical glide posture, showing the soft angles of the swept-back wings (as well as the very slender body and long tail). While boldly patterned, the bird appears pale overall. Utah, September JL

7. A male American Kestrel showing a configuration that would be typical on a bird flying in a casual or leisurely manner. You get a sense of buoyancy here, but the wandering, tacking, and rising and falling flight that is typical of migrating birds must be left to your imagination. Utah, October JL

8. A female American Kestrel in a soar. Note first the overall slender, almost delicate lines. Also evident, the slender banana-shaped wings and overall paleness that are so characteristic of this species. Cape May, October CS

9. A hovering female American Kestrel. This is how American Kestrel signs its name. Merlins don't hover. Utah, November JL

Merlin

The Merlin is to an American Kestrel what a Harley-Davidson motorcycle is to a bicycle. Superficially the two are similar. Both are small North American falcons that both perch-hunt and hunt on the wing and take a variety of prey. Given time for study, a Merlin might appear slightly larger but, except for the larger female Merlins, not dramatically so. When a Merlin takes flight, however, all similarities evaporate. In the air, the difference between a Merlin and an American Kestrel is not a matter of degrees, it is quantum.

The Merlin is a northern breeder, associated with trees only during nesting season. Even then, it shuns deep forests. It is a bird of the edge, nesting in proximity to burns, open meadows, muskeg—even open tundra, where it nests on the ground. The Prairie, or Richardson's, Merlin (subspecies *Falco columbarius richardsonii*) uses cottonwoods on the tree-poor prairies, appropriating the nests of crows or magpies (sometimes along residential streets, in parks, and in cemeteries of urban centers). In recent years this species has expanded its breeding range south and east, nesting in suburban Ithaca, New York, and on Martha's Vineyard, and its winter range has expanded into the northern Great Plains.

Outside the breeding season, Merlins resort to open spaces: barrier beaches, tidal flats, open range, and marshlands. The birds also reside in cities and small communities where they find all a Merlin could ask for: unlimited hunting perches; open, treeless avenues; and a surfeit of prey (primarily House Sparrows). Cemeteries can be favorite hunting and roosting places.

Three Merlin subspecies are found in North America, and all are distinguishable. The nominate and widespread Taiga Merlin, *F. c. columbarius,* has a breeding range that encompasses most of Canada and extends into several northern states. This is the subspecies most commonly seen at hawk watches in passage to and from its winter range—along both coasts and in open portions of the West, south into Central America.

The pale Prairie Merlin nests in the prairie regions of Canada and the United States and winters primarily in Oklahoma and Texas. The Black Merlin (*F. c. suckleyii*) breeds in the coastal forests of the Pacific Northwest and may, in winter, wander south to California and Nevada.

Merlins are feisty, pugnacious raptors with little tolerance for other birds of prey (including other Merlins). They will go out of their way to harass a bird that crosses into their territory or trespasses into their airspace. Since this species seems just as easily provoked during migration as at any other time of year, a Merlin's territory may be inferred to be wherever it happens to find itself.

IDENTIFICATION. Male Merlins are about the same size as American Kestrels. Large females are about the size of pigeons. Females and juveniles are brown above (females are slate brown, juveniles dark chocolate brown) with heavy streaking on the underparts that contrasts with a white throat and buffy undertail coverts. The underwings are checkerboard patterned. The net effect is a *small dark falcon.*

Caution: Western Taiga Merlins are paler than their eastern counterparts. Expect less contrast between Taiga Merlins and American Kestrels in the West.

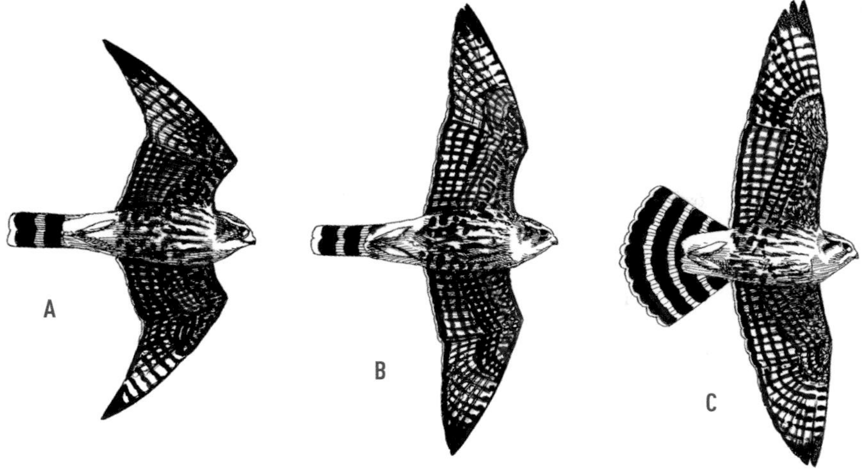

Merlin, underside. Juvenile **(A)**. Juvenile **(B)**. Adult male **(C)**. Small, dark, and angular, with relatively short, very pointed wings, fairly long tail, large head. In soar **(C)** wings are held as clean triangles. In all flight attitudes there is a slight step from the outermost secondary to the slightly shorter innermost primary. Plumage variation is minimal; all birds are dark, with checkered underwing, streaked body, black tail with fine white bands, and contrasting white throat and buffy undertail coverts. Variations include lightly streaked and pale-faced adult male **(C)** or heavily streaked, hooded immatures **(A, B)**, with adult females in between.

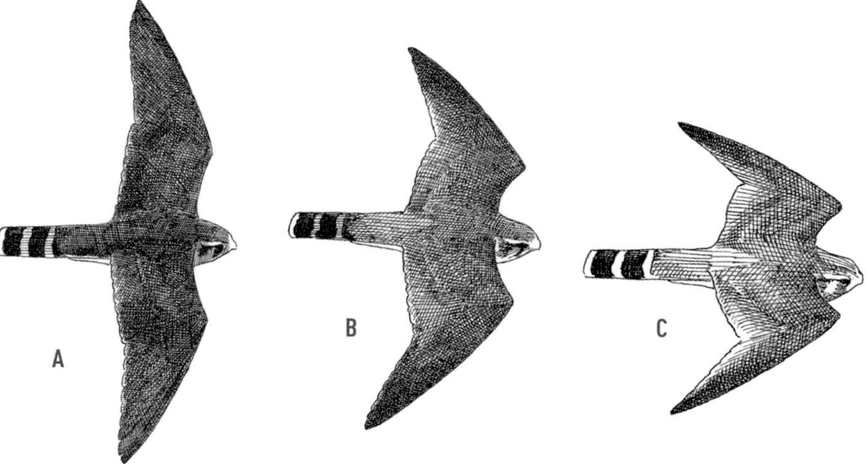

Merlin, upperside. Juvenile **(A)**. Adult male (*F. c. columbarius*) **(B)**. Adult male (*F. c. richardsonii*) **(C)**. Adult males blue above with pale rump, extremely pale powder blue in Prairie race *richardsonii*. Females have similar pattern in gray-brown; juveniles are uniformly dark chocolate brown.

Adult males are much paler overall: bright metallic blue above, rose blushed on the underwings, chest, and undertail coverts below. Streaking is finer, indistinct at a distance. Flight feathers on the underwings shine silver.

In all plumages, the celebrated falcon mustache mark is faint, smudgy, and indistinct. (In the case of adult male Prairie Merlin, the mustache may be absent; in the case of Black Merlin, it is generally lost to an all-dark hood.)

The Merlin is a stocky and compact falcon. The overall lines of the bird are stiff and angular. The body is chestier and heavier than that of kestrel. The tail is wider and thus appears shorter. The wings are longer than kestrels but broader, particularly in the arm, imparting a stocky appearance. The angles are harsher and sharper. In a full soar, the wings form isosceles triangles that meet the body flush and broad (that is, the wings don't pinch inward at the base). Wingtips are sharply pointed, not rounded.

At close range, on a fanned or folded tail, fine white lines set against a black backdrop are discernible. At greater distances, the lines disappear, and the tail appears dark. From below, the closed tail appears two-toned, half-dark and half-light — dark from the tip to just past the midpoint, light from just shy of the midpoint to the trailing edge of the wings. (*Note: The actual tail of Merlin is uniformly dark with pale narrow bands. It is the contrastingly pale undertail coverts that impart the two-toned, half-pale/half-dark effect.)*

Merlins, like kestrels, have a row of pale dots along the trailing edge of the wing that appear when the birds are backlit. But the dots are buffy and indistinct, never as bright and prominent as the string of pearls seen on kestrels.

IN FLIGHT. The flight of a Merlin is its most distinguishing characteristic. It is point to point, fast, and direct. It does not wander. Hunting and migrating birds frequently drop down, like a cruise missile, to hug vegetative or geographic contours. Wingbeats are quick, emphatic, continuous, regular, and powerful. They seem like short piston strokes, all the power concentrated in the flicking downstroke. The cadence and motion are not unlike the wingbeats of Rock Dove, but Merlin's wingbeats do not have the hesitant or halting quality that Rock Dove's sometimes show.

Merlins soar well but glide less frequently than kestrels. When a Merlin glides, the wings are angled down, not curved down at the tip like kestrels'. When Merlins set their wings, they start losing altitude *fast.* Gliding kestrels float and slow down; Merlins descend and maintain their speed but resort to powered flight quickly.

Because of their ground-hugging tendency, Merlins are frequently not seen approaching. On ridges, particularly, the small dark falcons are easily lost to the dark backdrop of trees below. Merlins appear suddenly, make a pass at the owl decoy, then continue on their way almost before onlookers can shout, "Merlin!"

Merlins are usually solitary birds (thanks to their bad disposition), and juveniles will frequently go out of their way to harass other birds in migration. An observer may use this aggressive tendency for identification purposes and as a means of detection. High-flying Merlins often betray themselves and distinguish themselves because they are vigorously harassing another raptor (even ones as large as Golden Eagle).

The attitude of other, nonraptorial birds can also serve as an oblique aid to Merlin identification. While some small to medium-sized passerines will ignore or even harass kestrels, they avoid Mer-

A Merlin chasing a Golden Eagle.

Merlin flapping: small, angular, and dark. White throat and buffy undertail coverts are often helpful in distinguishing this species from kestrel at a distance.

lins. For example, where a large number of Tree Swallows gather, an approaching kestrel will gather a small cloud of harassing swallows, while an approaching Merlin will bore a bird-free hole in a swallow-filled sky.

MERLIN SUBSPECIES. The preceding description was based principally upon the widespread, nominate race, *F. c. columbarius*. The two other North American Merlin subspecies are fundamentally similar but their plumages are distinctive enough to warrant separate treatment.

Prairie Merlins are larger, paler, and more finely marked than the nominate race. The female is pale gray-brown above and sparsely streaked with tan below. Male Prairie Merlins are pale, *powder blue* above, tan with fine rufous streaking below (for you photographers, it's a Taiga Merlin overexposed by two f-stops). Juveniles are similar to females: pale brown above, with pale streaks below.

The Black Merlin of the Pacific Northwest is slightly larger than the nominate race and much darker. Females are virtually black above and dark brown and heavily streaked below. Males are charcoal gray above and darker below than birds of the nominate race. The face is dark with the mustache mark often eclipsed. Tails may be all or mostly dark, with only some narrow and broken white banding. (Again, for you photographers, it's a Taiga Merlin underexposed by two f-stops).

1. A juvenile Merlin (females have very similar plumage). Note overall dark plumage (American Kestrel appears overall pale), checkerboard pattern of the underwings, and narrow white bands on the black tail. Also notice the indistinctly defined mustache. Cape May, October KK

2. A soaring and banking juvenile Merlin. Overall more robust, compact, and darker than American Kestrel. Note, again, the indistinct mustache. Cape May, October KK

3. An adult male Merlin or "Blue Jack" from above. In good light, the blue upperparts can be seen at considerable distances. But if the bird is approaching, distant views will be momentary. A Merlin's speed will ensure this. Cape May, October KK

4. An adult male Merlin in a fast glide. Note the acute angle on the wing and sharp-edged (not soft or curved) lines. Birds showing this wing configuration are traveling very fast and, unless using an updraft, losing altitude. Cape May, October KK

5. A juvenile Merlin in a fast glide. Overall dark plumage is broken by two pale points — a white throat and the basal half of the tail/undertail coverts. The "tail" of high-flying birds appears half-black, half-pale. (*Note: The pale portion is really the undertail coverts, not the tail, but projecting behind the wings as they do, they look like part of the tail.*) Cape May, October JL

6. A juvenile or adult female Merlin overhead in a fast glide. Just a small dark falcon. Not a kestrel because the too-broad wings are also too dark below and the distal half of the tail appears black with narrow pale bands. Not a Peregrine because the wings are too short and the tail is too narrow and straight. Peregrine tails are broader and commonly taper modestly toward the tip. Cape May, October CS

7. Female Richardson's or Prairie Merlins are overall paler than the more widely distributed Taiga Merlin. Also, not necessarily a field mark, but more than other falcons, Merlins just plain look ornery. Utah, October JL

8. An adult male Prairie Merlin in rapid direct flight. Compare the pale plumage to the Black Merlin at right. It is a bird of a different color! Arizona, December NH

9. A female Black or Pacific Coast Merlin in powered flight; it appears considerably darker than Taiga Merlin. The broad-based and acutely angled wings, shown here, are classic for all the races of Merlin. California, November TJ

10. This juvenile Cooper's Hawk (right) is learning how feisty and aggressive juvenile Merlins can be. In autumn, these journeymen Merlins will go out of their way to harass other hawks (and one another). This image shows well the Merlin's contrastingly pale throat and two-toned (half-dark, half-light) "tail." Cape May, October KK

Peregrine Falcon

The Peregrine Falcon, a medium-sized raptor, has long been a source of inspiration. During the Middle Ages, it denoted social status — only lords could fly a Peregrine from the fist. More recently, the bird served as an environmental rallying point, the poster child for the battle to ban the use of the insecticide DDT.

These considerations aside, the Peregrine is a creature whose mastery of flight sets new standards for the word *perfection.*

All three native North American subspecies are mostly cliff-nesting raptors. The southern race, *Falco peregrinus anatum,* once had two distinct populations: one in the East and one in the West. Widespread use of DDT in the years following World War II completely eliminated the eastern population and seriously depleted the population in the West. Luckily, following four decades of active management and restored natural productivity, populations have mostly recovered.

The northern Tundra Peregrine subspecies, *F. p. tundrius,* was likewise affected by DDT, its numbers reduced by half. With a breeding population found from Greenland to Alaska, and a wintering range that draws some individuals to the southernmost reaches of South America, this long-distance migrant is the Peregrine seen most frequently at most North American hawk-watch sites.

A third subspecies, the dark, coastal, Peale's Peregrine, breeds and winters from the Aleutians to the Queen Charlotte Islands. While it is recorded regularly in California, much of the population is nonmigratory.

A case could be made for classifying a forth subspecies of Peregrine in North America. After the extirpation of the eastern *anatum,* an active reintroduction program was initiated. Nestlings with genetic links to native and nonnative Peregrine subspecies were bred in captivity and released or "hacked" into both traditional and nontraditional sites. Many of these birds, and their descendants, established themselves in metropolitan areas, finding in cities a wealth of clifflike ledges and a surfeit of avian prey (including, most notably, Rock Pigeon). These urban Peregrines have become a widespread, entrenched, and largely welcome part of the urban avifauna. In the absence of any official classification, we suggest that the birds be called *F. p. urbanii,* the Urban Peregrine.

The Peregrine's diet is almost exclusively avian (although bats are not uncommon prey). Its most celebrated mode of hunting is from on high. Hunting Peregrines "tower up" over likely prey and engage their targets by employing a meteoric stoop (the actual speed of which remains a matter of considerable debate but may exceed 150 miles per hour). Peregrines are also adept perch hunters, searching for and then swooping upon prey from an elevated perch overlooking a river, lake, marsh, or beach.

IDENTIFICATION (applies to all subspecies). The Peregrine is a medium to large falcon with a wide range in size (and even shape) between the small male and the larger female. Adults and juveniles differ in plumage. Young birds have a brown back, heavily streaked underparts, and a more lightly streaked, buff-colored chest and throat. Adults are dark blue-gray above (head slightly darker) and heavily to

Peregrine Falcon, underside. Juvenile **(A)**. Adult **(B)**. Long winged and fairly long tailed. In soar, wings are held straight with evenly curved edges, tail fanned broadly; wings may appear blunt tipped in soar but at all other times are sharply pointed. Males appear relatively small, slim, and buoyant; the largest females are broad, heavy, and powerful. Adults are light chested, with fine barring on belly, and have heavy black mustache; juveniles are streaked on body (some heavily, some lightly) and have distinct brown mustache, sometimes very thin in pale *tundrius* birds. In all plumages the wings and tail are dark and clean edged.

faintly barred below, with a gleaming white (to tawny or pinkish) chest and throat. Adult males are whiter below than females.

Adults and juveniles have a distinct bold mustache stripe. In combination with the dark head, it makes the bird look as though it is wearing a gladiator's helmet.

IN FLIGHT. The Peregrine is an *extremely* long-winged and fairly long-tailed falcon. Juvenile birds have a longer tail than adults. The body, particularly the chest, is heavy and broad. The tail, when closed, seems broad despite the length, but gradually tapers toward the blunt tip.

A soaring bird has wings that taper both fore and aft. They resemble long narrow candlesticks, or an elongated lancet arch. The tail, when fully fanned, may form a semicircle that almost touches the trailing edge of the wing.

A silhouette of a soaring Peregrine is distinctive, almost unique. It should be confused only with another large falcon or Mississippi Kite.

In a soar, the wings are held perpendicular to the body and straight (not drooped or elevated). Turns are wide and slow and have a dreamlike quality. When a soaring bird turns or banks, the hand lifts stiffly upward, the wing momentari-

Adult Peregrine flapping.

ly bent in a one-winged dihedral. In a glide, the wings droop ever so gently along their length.

In powered flight, the wingbeat is continuous, fluid, rhythmic, undulating, and whiplike. The flap appears shallow, the energy pulsing or rippling down the wing. Wingbeats seems less a continuous series of individual flaps than a run-on movement with one flap finishing as the next begins. (*Note:* The whipping wingtips of Peregrine actually navigate a deeper arc than the rest of the wing—as frozen images show. The rapid movement and diminished mass cause observers not to notice the wingtips, but instead to focus upon the balance of the wing, with its shallower range of motion.)

In cadence and execution, the wingbeat is not unlike that of Common Loon and also recalls Double-crested Cormorant. *Caution: The shallow wingbeat, which is a hallmark of Peregrines in level flight, becomes considerably deeper when the birds are just leaving a perch or accelerating to pursue prey.*

Despite the comparative similarities just offered, the wingbeat of a Peregrine is very distinctive and, to those familiar with it, as identifiable as a fingerprint *no matter what the distance.*

The usually solitary adults often migrate in pairs, and juvenile trios are not unheard of.

Curiously (and excepting Mississippi Kite), the bird most often confused with Peregrine is not even a falcon. Seen gliding or soaring high overhead, Northern Harrier is similar in size and proportions to Peregrine. But the wings of Northern Harrier do not taper, and the tail is proportionately longer than the tail of even a juvenile Peregrine. Any residual confusion should last no longer than the first series of wingbeats. The rapid, fluid, steady rippling wingbeats of Peregrine could hardly be more different from the lazy, loping wingbeats punctuated with a floating glide that characterizes the flight of Northern Harrier.

At times, a soaring Peregrine can also bear uncanny likeness to a soaring Broad-winged Hawk, a buteo! When fully fanned, the broad tail of Peregrine effectively masks the extreme length of the Peregrine wing, creating a silhouette that is very like the candle-flame-winged Broad-winged Hawk.

Confusion should be only momentary. The differences between buteos and falcons are considerable, and they will manifest themselves given a little study.

PEREGRINE SUBSPECIES. The adult *anatum* Peregrine is slate gray to black above and heavily barred below, except for a whitish upper chest and throat. The lower chest and belly commonly bear a rufous wash so that the bird has a subtle orange or pinkish cast when seen from below. The dark upperparts extend, hoodlike, onto the head and face. The black mustache stripe is sharply defined: bold, broad, and obvious. Juvenile birds are generally dark brown above, heavily streaked below, with the underparts on some individuals showing a rusty wash.

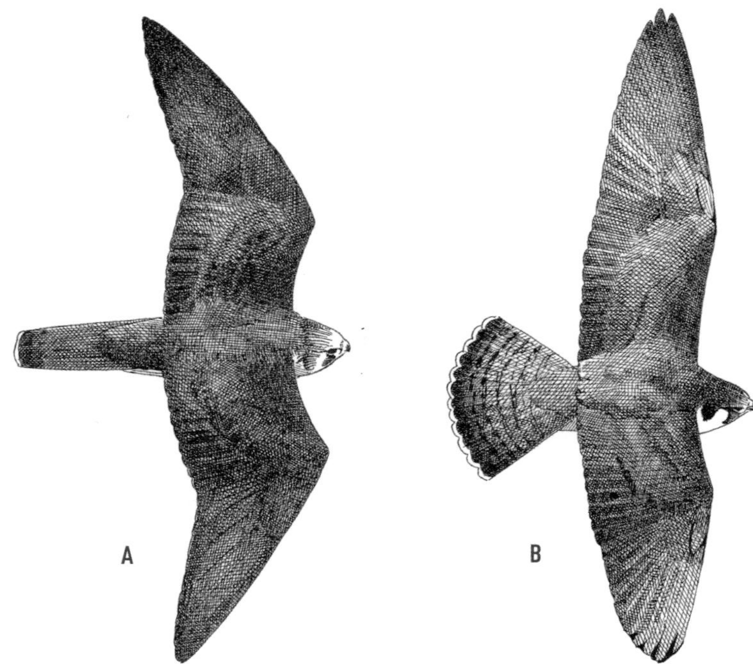

Peregrine Falcon, upperside. Juvenile (*F. p. tundrius*) **(A)**. Adult **(B)**. Adults appear more or less uniformly dark slaty blue above, palest on rump (males paler and bluer than females), with dark wingtips, tail tip, and head markings. Juvenile is uniformly dark gray-brown except for some fine light edges on scapulars and pale buffy crown on *tundrius*, as illustrated.

The Tundra Peregrine is a comparatively paler bird. Adults are slate gray (a shade or two lighter than *anatum*). Males show a blue-gray cast to the back, wings, and tail. The dark head and mustache are more restricted than on *anatum*. Birds appear less hooded and more helmeted. The plumage on juvenile Tundra Peregrines is variable, ranging from brown to tan. Underparts are tawny and lightly streaked, and lack the rusty wash found on *anatum*. Many individuals have a distinct blond crown, an indistinct helmet, and a mustache mark that is narrow and not particularly prominent. They can fool you into thinking they are Prairie Falcons until you see the underwings.

Larger on average than other Peregrines, adult Peale's Peregrines are dark gray with a black hood — generally darker than *anatum* (with more spotting on the breast) but whiter on the face, throat, and (beneath the heavy and extensive barring) underparts. Peale's lack the rusty wash often found on *anatum*, and so appear crisper, displaying more contrast.

Juvenile Peale's Peregrines are very large and *very* dark — charcoal brown above, dark and heavily streaked below. Like *anatum*, the head is hooded more than helmeted — in fact, the mustache may cover the face. Because of its size and very dark plumage, the juvenile Peale's Peregrine might easily be confused with a dark-morph Gyrfalcon.

1. An adult Peregrine Falcon. Adults show a whitish breast (males more than females), barred underparts, and, typically, a distinct mustache slash that creases the face. It's certainly a long-winged falcon, but also long tailed. The extreme length of the wings often masks or diminishes tail length. Cape May, September KK

2. The dorsal view of an adult Peregrine, banking in a glide. In general, birds appear very neat, trim, and clean lined. Remember that many Peregrines, particularly males of the Tundra race, are paler gray above than this individual. Alaska, July TS

3. A robust adult Peregrine female, showing a heavy body and broad wings. Alaska, July TS

3. An adult Peregrine in unhurried direct flight. Even at a seemingly leisurely pace, they still measure distance by the horizon and cover vast areas with ease. New Jersey, December CS

5. A juvenile Peregrine banking. Juveniles are brown (not slate blue) above, buffy (not whitish) below, and spotted or streaked, not barred. And while the mustache is less crisply defined, it is still bold and prominent. The touch of gold on the crown distinguishes this as the Tundra subspecies that breeds in the Arctic. Cape May, October KK

6. A juvenile Peregrine (same bird as #5) in a fast glide. Both this and #5 are classic Peregrine profiles. Commit them to memory. KK

7. A soaring juvenile Peregrine from above. Note the very fresh plumage and spotting on the tail. Some Tundra Peregrines are fully blond headed, even more so than this individual. Utah, October JL

8. A somewhat pale juvenile Peregrine. Expect a great deal of variation between individual young birds. Note the uniformity of the underwing. The pattern seems uniform from underwing coverts to flight feathers (that is, no contrastingly paler flight feathers as on Gyrfalcon). Cape May, October JL

9. A juvenile Peregrine in a full soar. The profile can be surprisingly kestrel-like (given the blunted wingtips, long narrow wings, and long tail), or even somewhat gull-like. Also, on high-flying birds, the fully fanned tail masks the length of the wing, making a high-soaring Peregrine appear, at times, to be shaped like a Broad-winged Hawk. Cape May, October CS

10. A juvenile Peregrine, gliding high overhead. Another archetypical profile that shows very long wings and long tail tapering very slightly toward the tip. Note, again, the uniform underwings. Compare this photo to Merlin #6. Cape May, October CS

11. LEFT: One of nature's most impressive displays: an adult Peregrine in a full stoop. It doesn't get any better than this. Alaska, July TS

12. An adult *anatum* Peregrine. All previous birds have been Tundra Peregrines. Western *anatum* birds show a buffy-peachy wash and a heavy hood; backs and upperwing coverts are dark slate-gray compared to the paler bluish gray wash of Tundra. Arizona, December NH

13. A smaller male *anatum* Peregrine on the left, a larger female on the right (Gila Woodpecker in the male's talons). The wingbeat of Peregrine in level flight is shallow, fluid, rippling, or undulating and even. You can conjure this flex in the left-hand bird. Birds in fast pursuit use deeper wingbeats, like the bird at right as well as #12 above (also see p. 325). Arizona, November NH

Prairie Falcon

If you reside in the western half of the United States, then you are no doubt familiar with this butte-haunting ghost of a falcon, the "prairie wraith." The bird is a spirit in league with short-grass prairie, sandstone cliffs, open sage, alpine tundra, and agricultural cropland. In such a habitat, your chances of seeing the bird are very good. Beyond a range that extends roughly from the American steppes westward to the coast, north to southern portions of the Canadian Prairie Provinces, and south to central Mexico, your chances of seeing North America's desert falcon are considerably diminished (but better than spotting a real ghost). After the breeding season, the birds do exhibit an eastern pattern of distribution.

Prairie Falcons are most often likened to a pale Peregrine, an understandable if not terribly apt comparison. Still, it is convenient to become familiar with a thing by measuring it against a known quantity. The Peregrine's greater distribution and mobility make it the more familiar of the large falcons over most of North America, hence the comparison. But the Prairie Falcon is actually more closely allied to the Gyrfalcon, another of the desert falcon group that includes both the Gyrfalcon (*F. rusticolus*) and the Saker (*F. cherrug*) and Lanner (*F. biarmicus*) falcons of the Old World.

While Peregrines are bird specialists, Prairie Falcons target small mammals, as well as many bird species of open country. From a strategic power-line tower, from a sandstone ledge, or from the air, the bird marks its prey and flies rapidly along a route calculated to conceal its approach for as long as possible. The flight is usually low and hugs the contours of the ground.

The falcon also uses a modification of the same low, coursing flight to locate and surprise prey — a powered patrol whose objective is to catch ground squirrels unaware or unnerve prairie birds into flight. Prairie Falcons also tower and stoop in the manner of a Peregrine.

Not surprisingly, in view of the bird's hunting style, the Prairie Falcon is usually sighted when it is traveling from one place to another. Also understandably, most sightings are of brief duration; the bird is either coming (if you are lucky) or going (if you are not).

Prairie Falcons depart from the well-developed migratory tendencies of the previous three falcon species. More nomads than migrants, in winter Prairie Falcons expand their geographic range rather than abandon it.

In midsummer, after their young have fledged, the birds commonly leave summer-parched breeding areas, dispersing, in many cases, *up* to higher elevations, where prey is more abundant. Beginning in August, the birds vacate the high country and apportion themselves where prey is found — some west to the coast, some east to the edge of the Great Plains, some south, deeper into Mexico. A very few Prairie Falcons occur, as vagrants, east of the Mississippi.

IDENTIFICATION. The Prairie Falcon is the size of a Peregrine and, while slightly leaner, is similar in shape and proportions. Adults are pale brown above — distinctly paler than a typical juvenile Peregrine — and cream-colored below, with fine spot-

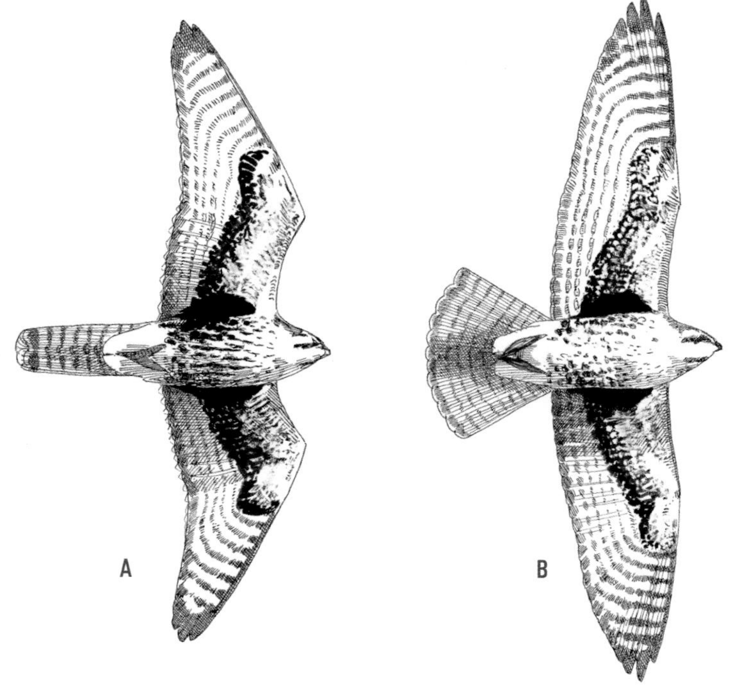

Prairie Falcon, underside. Juvenile **(A)**. Adult **(B)**. Slightly slimmer and longer tailed than Peregrine, and wingtips slightly more rounded. Best identified by plumage, much paler than on Peregrine; all flight feathers translucent with light gray markings. Body mostly white (adults) or buffy (juveniles), although the most heavily marked individuals can show a complete blackish belly band; face markings narrow and brown. Most of the underwing coverts are spotted and barred black; the resulting black bar along the underwing is conspicuous and diagnostic. There is little plumage variation, but juveniles are buffier, with more streaking below.

ting on the chest and belly. Juveniles are slightly darker, but still paler than a Peregrine. Underparts are buffy to whitish, more streaked or beaded than spotted, but overall more lightly marked than a Peregrine. Males and females differ only in size.

In the hawk-watching arena, adult and juvenile Prairie Falcon plumages are functionally identical. In essence, they are *large pale falcons.* Once you get the size right, and once you establish that a distant, soaring falcon has the color quality of a mirage, you hardly need to note any other defining field marks to know you are looking at a Prairie Falcon. But in case the mirage starts to flicker, here are some supportive field marks.

IN FLIGHT. In flight, the tail of Prairie Falcons (adults in particular) is paler and grayer than, and contrasts with, the darker upperparts. Peregrines show no such contrast. The facial pattern of Prairie Falcon, with its eye stripe and pale cheeks (even a pale collar), more closely resembles that of Merlin than the hooded or helmeted Peregrine. The Prairie Falcon mustache is narrow (not Peregrine-wide) and crisp (not Merlin-smudged) and the white cheek comes right up to the eye.

The Prairie Falcon's overall paleness accentuates the bird's next most conspicuous plumage characteristic: the bold dark-onto-black triangles that define the inner core of the bird's otherwise pale underwings. These dark triangles, beginning at the axillaries and encompassing much and sometimes most of the underwing linings, look like two broad black support struts beneath the wings.

These dark support struts are visible at tremendous distances. When the bird is high and light conditions harsh, they may indeed be the only thing visible as this mirage-colored bird melts into the sky. No other North American falcon has an underwing pattern like this. However, Ferruginous Hawk, another pale-bodied, pointy-winged denizen of the arid and open West, may mimic it.

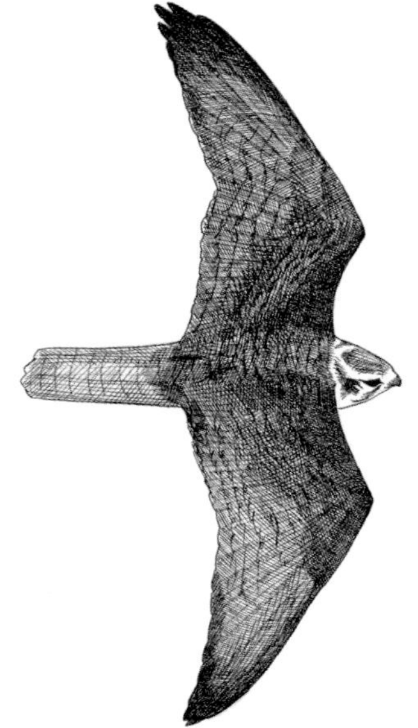
Prairie Falcon, upperside. Adult and juvenile are similar: always brownish, darkest on primaries, paler on rump and tail (outer tail feathers palest). Juvenile Peregrine is also brownish above but darker and more uniform, with a slaty cast.

Although structurally similar to Peregrines, Prairie Falcons seem less elegant and more slender. The head is prominent and square, the tail long but broad and blunt. The wings are long but less tapered and more uniformly wide along their length. The wingtips are rounder, blunter, a bit raggedy along the trailing edge. All in all, it is a falcon that is not quite a gentleman.

Prairie Falcons prefer gliding and sailing to soaring. As a result, they are commonly seen with hands pulled slightly back. In this attitude, they are very similar to Peregrine, but the wings may appear more curved than angled back, and at high altitudes the birds appear headless.

It is only when the birds engage in a full soar that the blunter wingtips of Prairie Falcon become evident, and *now* the shape really recalls the gentle angles of American Kestrel (in fact, more than one observer has likened Prairie Falcon to an overgrown kestrel).

In powered flight, the tail may narrow to an acute point. A Peregrine's tail tapers but always remains blunt. In flight, the rhythm and cadence of the Prairie Falcon are like those of a Peregrine but the movement and its mechanics differ slightly. The wingbeat of a Peregrine is fluid, undulating — a ripple of motion that flows down the wing. The wingbeat of a Prairie Falcon is stiffer, choppier, more mechanical. The motion seems more centered at midwing. The wrist moves in short, stiff, up-and-down strokes. The wingtips move in counterpoint.

One last point, but a significant one: like kestrels, Prairie Falcons hover. Not habitually (like American Kestrel), nor at length, but occasionally. Peregrines do not.

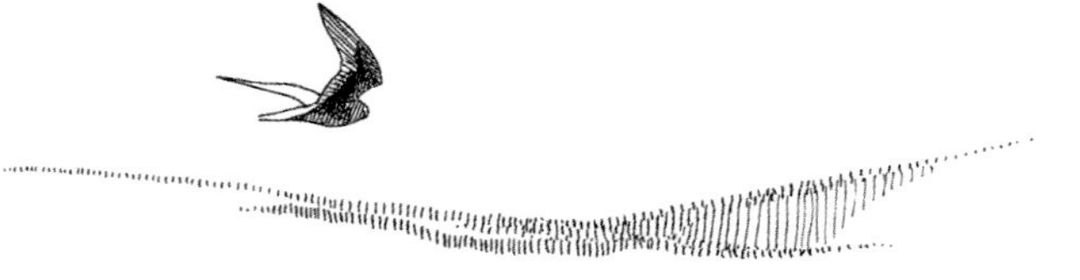
Typical view of Prairie Falcon, flying fast and low.

1. An adult Prairie Falcon in a glide. The desert falcon of the American West. About the same size and shape as Peregrine, but overall paler and grayish brown. Mustache is sparser. The body of adults is lightly spotted. Underwings sport dark wingpits and coverts that run along the wing like a dark support strut. Utah, April JL

2. An adult Prairie Falcon gliding on an updraft. The angles of the wings are softer, often blunter tipped, more kestrel-like than that of Peregrine. The long tail is typically broad, straight edged or slightly fanned. But it's the dark wing lining set against the pale underparts that really stands out. Head-on, birds appear pale headed. Utah, April JL

3. A particularly pale adult Prairie Falcon in a glide. The flight feathers on backlit birds appear translucent, contrasting with the dark wing lining. The wings of Peregrines are monotone and opaque, and lack contrast. Utah, May JL

4. An adult Prairie Falcon in powered flight — this is a fairly heavily marked individual. In addition to softer angles on the wings, Prairie Falcon often seems not quite as clean-cut or trim as Peregrine, more unkempt or rough-cut. Prairie Falcon's wingbeats are somewhat stiffer, not as rippling or fluid as Peregrine Falcon's. Note: the golden cast is caused by morning or evening sunlight. Utah, October JL

5. The dorsal view of an adult. Compare this view to Peregrine #7. It is conspicuously paler than Peregrine, and with a paler tail that contrasts with the body and upperwings. Utah, May JL

6. A backlit juvenile in a full soar. Compare this photo to Peregrine #9. It is lightly streaked below (not spotted as adult) but still overall pale and showing the dark support strut along the underwing. Also, compared to Peregrine, Prairie Falcon appears slightly more slender winged, making the head appear larger and, at times, more squarish. Nevada, October JL

Gyrfalcon

Here is a falcon that combines the size of a Red-tailed Hawk with the flight prowess of a Merlin, then adds an element beyond the measure of both. The largest of the falcons, and a true Arctic falcon, Gyrfalcon adults are presumed to remain on or close to breeding territory in the Arctic as winter closes over the land. In winter, as in summer, Rock and Willow ptarmigan are the dietary mainstays, although Gyrfalcons are capable of taking almost any size avian prey.

In winter, some adult females may relocate to coastal areas that offer more hunting opportunities and some portion of the population winters out on the Arctic ice sheet. Young birds, particularly juvenile females, may wander great distances. One radio-tracked bird amused herself by wandering back and forth between Kodiak Island, the west coast of Alaska, and western Siberia for much of one winter. These peregrinations sometimes carry Gyrs into southern Canada and northern border states. Their discovery invariably results in a concentration of eager hawk watchers hoping for a glimpse of this, the ultimate falcon.

The size of the bird and its broad proportions sometimes make Gyrfalcons look deceptively slow. They are not. A Peregrine, for all its celebrated speed, must commonly rely on a meteoric stoop to capture prey. Gyrfalcons simply fly prey down — overtaking prey in flight or driving it to the ground. Grounding does not grant prey the immunity it might gain if pursued by a Peregrine (an aerial specialist). Gyrfalcons are as tenacious on the ground as they are accomplished in the air.

Three color morphs are generally recognized: white, dark (gray/brown), and gray (the intermediate plumage with the widest distribution across North America). There is much plumage variation between types, and the idea of distinct morphs is more a matter of convenience than of biological or geographic significance.

IDENTIFICATION. The Gyrfalcon is a large robust falcon as big as, or larger than, most female Peregrines. Some male Gyrfalcons may be smaller than some female Peregrines, particularly the larger, darker Peale's Peregrine of the Pacific Northwest. Some female Gyrfalcons are larger than Rough-legged Hawks and may weigh twice as much as the average Peregrine.

White-morph adults are white overall, with some dark flecking on the upperparts and little, if any, patterning below. The face is unmarked (although the shadowy suggestion of an eye line or mustache may be visible). Wingtips are dark, both above and below.

Gray-morph adults have a gray back with lighter edges on the feathers (creating a scaled effect on wings and backs). Underparts are white with dark flecking on the underwing coverts and body. An indistinct mustache that becomes more prominent near the chin is evident.

Dark-morph birds are smoky brown above and very heavily spotted or streaked below — some birds are wholly dark. The mustache mark may be eclipsed by an all-dark hood.

Juveniles in all three morphs show the same patterns, but they are darker and more heavily patterned than adults (and juveniles are the birds most likely to appear

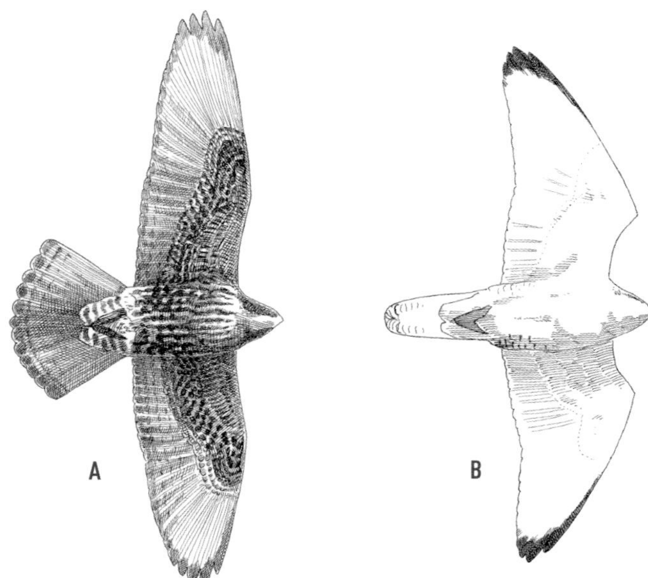

Gyrfalcon, underside. Dark morph **(A)**. White morph **(B)**. Large, broad, heavy bodied; wings unfalconlike when soaring, broad and distinctly rounded; tail long and broad; head large. White morph unmistakable (but beware of albinos of other species, particularly Red-taileds, which are more likely than white Gyrfalcons to occur in, say, Alabama); white plumage relieved only by black tips on primaries and fine black barring on flanks. Dark morph distinguished from Peregrine and Northern Goshawk by dark underwing coverts, contrasting with paler gray flight feathers; by generally muted, pale pattern without sharply defined dark markings; and by shape. Juvenile dark morph (shown here) has streaked pattern below; adults more spotted and barred.

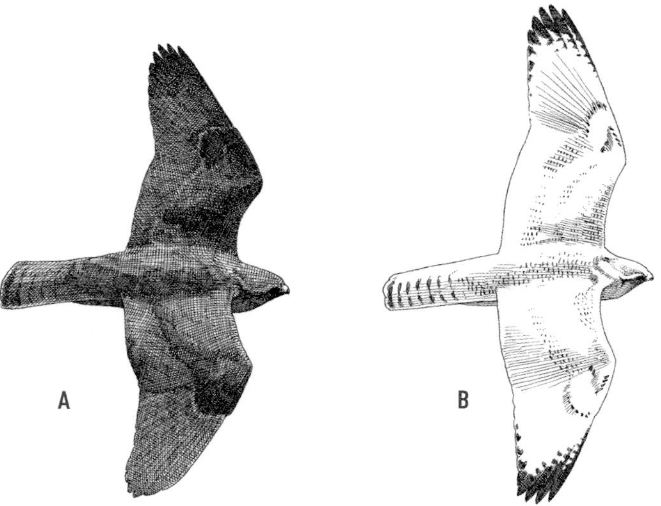

Gyrfalcon, upperside. Dark morph **(A)**. White morph **(B)**. In characteristic flight posture this falcon exhibits hunched shoulders, drooping head and tail, broad back, crooked wings. White birds are more heavily marked above than below. Bird shown in **(B)** is near the light extreme, with only faint barring on back and wing coverts; others can be evenly barred from crown to rump and appear gray; all have black wingtips. Dark birds may look spotted or uniform, all in various shades of smoky gray or brown.

south of the birds' breeding range). *On juvenile dark and gray morphs, the primaries and secondaries are paler than the underwing coverts, and somewhat translucent — a characteristic not shared with Peregrines.*

IN FLIGHT. This is a large, robust, humpbacked, heavy-chested, potbellied falcon. The wings are long and broad — broad in the arm *and* the hand with a shape more reminiscent of Prairie Falcon, sharing with Prairie Falcon blunter, rounder wingtips than Peregrine.

The tail appears disproportionately long and very wide. When the tail is seen from below, its length is masked by its breadth. When the bird is viewed from the side, the tail's distinctive length is apparent. At times, in direct flight (and like Prairie Falcon) the tail may taper to an acute point (a characteristic never shown by Peregrine).

In flight, it is *eminently* falconlike — direct, strong, fast. Wingbeats are rapid, steady, and shallow, slightly slower and not as whippy or undulating as the wingbeats of Peregrine. At times, the motion of the wing appears limited to the hands. Energy doesn't flow down the wing so much as it seems concentrated toward the tip; and while the wings seem hardly to be moving, the bird nevertheless covers ground incredibly fast.

Perhaps because of size and plumage, a bird that Gyrfalcon is sometimes confused with is Northern Goshawk. *Rule of thumb: If the gray, humpbacked, potbellied, broad- and pointy-winged bird you are undecided about disappears into the trees, it isn't a Gyrfalcon.* Gyrfalcons are birds of open country. Goshawks are forest birds.

1. A gray-morph adult Gyrfalcon, showing a robust body; long, wide, somewhat blunt-tipped wings; and long, often somewhat tapered tail. The mustache on this bird, while fairly prominent, might suggest Prairie Falcon to some, but note the color and the pale wing linings. The right wing shows a hint of translucency, which Peregrines do not show. Idaho, November JL

2. A gray-morph adult Gyrfalcon in a glide (and in heavy molt). Note the long broad tail and translucency along the flight feathers of the wing. Alaska, June TS

3. A gray-morph adult Gyrfalcon (same bird as #2) in direct, active flight. The wing motion seems centralized in the outer portion of the wing (i.e., in the hand). TS

4. A Gyrfalcon in a glide. Note the robust body; long, broad, but blunt-tipped wings; long, narrow, tapering tail. Note, too, the translucent flight feathers. Backlit and seen in silhouette here, this is actually a white-morph adult male. Pennsylvania, January CS

5. A dark-morph juvenile Gyrfalcon (left) soaring and banking with a juvenile Great Black-backed Gull. Some dark morphs are dark gray, some brown like this individual. Note the pale flight feathers contrasting with dark underwing lining. This individual (a male) is about the same size as, and has plumage characteristics of, a dark juvenile Peregrine, but differs in its broad, blunt-tipped wings and contrastingly paler flight feathers. Connecticut, January CS

6. A white-morph Gyrfalcon in a fast glide . The tail on this bird appears uniformly wide, but it is changeable, sometimes widening toward the tip, sometimes tapering. Alaska, June TS

7. A white-morph Gyrfalcon in a full soar (same bird as #6). Note the translucent flight feathers and the broad, blunt-tipped wings (blunt for a falcon). TS

8. The dorsal view of a banking white-morph Gyrfalcon.Same bird as #6 and #7. TS

9. A gray-morph juvenile Gyrfalcon in fast, powered flight. This is a paler bird and more typical than #5. In this view, notice the rounded, blunt wings; the broad tail; and the hint of a mustache pattern. Remember that in this plumage, Gyrfalcon can resemble Northern Goshawk in many ways. New York, October LO

PUTTING IT TOGETHER: Telling Falcons Apart

Falcon identification, unlike accipiter identification, is not a hairsplitting feat. Each falcon has distinct differences in plumage, shape, and flight. These distinctions notwithstanding, falcons may be more complicated to identify than accipiters. With accipiters, identification usually boils down to a simple choice: Sharp-shinned or Cooper's? Cooper's Hawk or goshawk?

Kestrels, Merlins, Peregrines, and Prairie Falcons share a battery of traits. Any one may be confused with another. Gyrfalcon is seldom seen at most hawkwatch locations. Should a dark morph appear, confusion with juvenile Peregrine is likely.

Gyrfalcon (left) soaring with Red-tailed Hawk. Size and bulk are similar, but Gyrfalcon is longer tailed and has more-tapered wings.

An approaching falcon has a distinctive wingbeat and manner of flight. Wingbeats are usually steady and uninterrupted. Buteos, accipiters, harriers, and kites flap in a series and glide. When falcons do glide, most lose altitude quickly. The lighter, more buoyant kestrel is the exception. Unless fighting a headwind, kestrels commonly flap and glide. When gliding they slow, but they don't sink.

Each species has its own distinctive cadence or manner of flight. Even when distance makes things like size, color, and shape impossible to determine, the manner of a falcon's flight can often identify the species.

An approaching Peregrine will look very long in the wing. The movement of the wing is fluid, rhythmic, undulating, and more shallow than deep. The wingbeat of Prairie Falcon is slightly stiffer, more mechanical than Peregrine's. It has the same cadence but lacks the fluid quality (the difference between a sprinter and a speed skater). The wingbeat of Gyrfalcon is slightly slower, more shallow and less undulating than Peregrines. Often the motion seems concentrated at the tip of the wing.

If you are fortunate enough to see an approaching Merlin (because Merlins usually appear without warning), you will see a small, dark (adult females and immatures), short-winged object moving at astonishing speed. Adult male Merlins are paler but still Merlin-fast — a flickering flame of a falcon. The wingbeats are rapid, constant, downward-flicking piston strokes. The flight is direct and steady, but the bird may drop a few feet to hug the contour of a hedgerow, a tree line, or the crest of a ridge. Merlins also jag quickly, left or right, to snatch insects in flight (especially in fall when dragonflies are migrating and many Merlins are hungry birds of the year) or go out of their way to harass other birds of prey. *Rule of thumb: If a falcon passes a perched raptor and does not go out of its way to take a shot at it, then it probably isn't a Merlin.*

A distant kestrel will appear small, slim, pale, and long winged. The wingbeats will be stiff and fluttery, with lots of up-and-down motion. On light-wind days, the

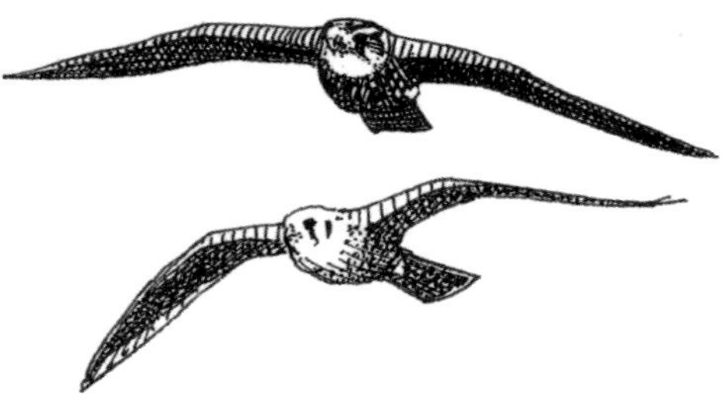

A Kestrel (bottom) and a Merlin approaching. Note the stiffer, more powerful appearance of the Merlin, as well as its darker color.

wingbeats of a kestrel may be punctuated by a glide on slightly down-drooped or wilted wings (like the classic accipiter flap, flap, flap, glide — except kestrels more commonly go flap, flap, flap, *flap, flap, flap, flap, flap, flap,* followed by a long, slow, floating glide).

A kestrel's flight is buoyant and wandering; the bird often fails to keep an even keel. But in a strong headwind, a kestrel will pull in its wings and pump strongly. In this situation, a kestrel is *more* like a Merlin.

The overhead silhouette of a Peregrine moving full speed ahead can be confusing. A juvenile male, with its wings pulled in, beating steadily, may resemble a large female Merlin. The flowing motion of a Peregrine's wings is not evident from below, but the wings will be longer and the tail noticeably broader than a Merlin's. In level, powered flight, a Peregrine's tail will also taper modestly toward the tip. A Merlin's tail does not taper and always appears narrower.

Be attentive to contrasts and patterns. From a distance, a Prairie Falcon will generally appear pale overall, and the contrast between a paler tail and darker upperparts is easily noted. From below, the dark support struts running along the underwing are prominent and apparent even at distances when the outline of the bird is not.

A female or juvenile Merlin will appear dark except for a hint of whiteness about the throat and a two-toned undertail (half-dark, half-light). A Peregrine (particularly an adult) will look very white chested, a helpful feature since adults have shorter wings and are, therefore, more Merlin-like in shape.

Going by, viewed wing-on, a Merlin loses none of its bullet gestalt. Females and juveniles, even at a distance, seem overall dark. Kestrels appear overall paler, particularly on the underparts, and females often show a hint of rufous in the plumage. From below, the Merlin's bold underwing checkerboard pattern is prominent. A male kestrel displays a row of white dots on the trailing edge of the wing, while female kestrels have buff-colored dots. Merlins may show this pattern, but it is muted and generally difficult to see.

Caution: Although the "Merlins are dark/kestrels are light" rule works well in fall (when many young Merlins are migrating), it does not apply to adult male Merlins. Adult male Merlins appear kestrel-pale at a distance. The blue upperparts are usually very apparent (and very un-kestrel-like). The two-toned undertail pattern on the Merlin, and the trailing row of lights on the wing of the kestrel, also help distinguish adult male Merlins and kestrels, as does, of course, behavior.

Adult male Merlins seem particularly stubby. Kestrels are slimmer and longer limbed. Kestrel wingbeats are hesitant and tentative, not Merlin-emphatic. A kestrel pumping into a strong headwind may recall a Merlin but, because of its inferior weight, a kestrel simply cannot buck the wind as well. It will fly more slowly than its larger, heavier relative and get pushed around.

When soaring, a kestrel has long, narrow, softly curved wings that resemble a long round-tipped candle flame. The tail appears long and narrow. A soaring Merlin has wings that look like broad, sharply cut isosceles triangles. The tail is short and not slim.

A Peregrine has wings that look like long, graceful, tapered candlesticks. The closed tail appears long, broad, and tapered toward the tip. When this species is gliding and soaring, the overall silhouette resembles a crossbow.

Prairie Falcons cruise and glide more often than they soar, but even when gliding and soaring, they often angle the wingtips slightly back. Fully extended, the tips are rounded, the hand and arm narrow and symmetrical, not elegantly tapered. It is not uncommon for high-soaring Prairie Falcons to be confused with American Kestrels (the falcon with "soft angles"). But if you are close enough to see the bird, you are almost certainly close enough to see the very un-kestrel-like dark wing struts running along the Prairie Falcon underwing.

Pointed-winged Kites

The Wind Given Form

SPECIES
Mississippi Kite, *Ictinia mississippiensis*
Swallow-tailed Kite, *Elanoides forficatus*
White-tailed Kite, *Elanus leucurus*

In Europe, Asia, and Africa, the term *kite* conjures the image of a large, dark, carrion-eating bird. In the Western Hemisphere, it most often applies to a small, graceful, aerial predator of insects, small reptiles, and mammals. The term probably comes from the Indo-European root *skut,* meaning "to go swiftly" (an apt description of the way kites swoop on their prey), although in current use the word describes a way of hanging motionless in the air — a trick that White-tailed Kite (as well as several buteo species) can perform.

Kites demonstrate a mastery of flight that makes most other birds, even other raptors, look tethered. Buoyant almost to the point of weightlessness, micromanaging the air with wings and tail, the birds swoop, stall, pivot, and fall like animate thistledown — or the wind given form.

If a sonnet were to be wrapped in feathers, it would look like a kite. If a high-flying dragonfly or a snake foraging in the canopy were asked to write the verse, it would read like an epitaph. Whatever their grace in the air, and however benign their facial expression, few predators are as artfully adept in their capture of prey as are kites.

North America's family of kites is divided into two groups. There are the agile, aerial specialists with pointed wings and wide distribution (Mississippi, Swallow-tailed, and White-tailed kites) — the three species discussed in this chapter. And there are the sedentary, paddle-winged kites of tropical and subtropical regions with limited distribution in the United States (Hook-billed Kite and Snail Kite) — discussed elsewhere in these pages.

Mississippi, Swallow-tailed, and White-tailed kites display remarkable energy and spirit and are gregarious, even during the breeding season. At no time is this

LEFT: An adult Mississippi Kite in a full soar. Kites are buoyant and acrobatic, consummate aerialists that are the most graceful of hawks. NH

group affinity more apparent than at the onset of and during migration. In August, in southern Florida, roost counts of staging Swallow-tailed Kites may exceed 1,000 individuals, and in southern Texas, migrating flocks in excess of 1,000 Mississippi Kites have been tallied. At Veracruz, Mexico, season totals for Mississippi Kites approach 300,000. White-tailed Kite, the most stay-at-home member of the trio, forms communal roosts.

In addition to their aerial finesse, pointed-winged kites are celebrated for their protracted soaring. When hunting, Mississippi and Swallow-tailed kites may be on the wing for hours at a time. They soar to great heights and can be seen as tiny dots against the backdrop of cumulus clouds.

Although the range of White-tailed Kite has expanded north into Washington State, and Mississippi Kite has recently been recorded breeding in Ohio, Connecticut, and New Hampshire, kites are basically southerners. The entire United States population of Mississippi and Swallow-tailed kites withdraws to Central and South America shortly after the end of the breeding season. Its pioneering efforts northward notwithstanding, White-tailed Kite is widely distributed as a year-round resident in Central and South America.

In North America, Mississippi and Swallow-tailed kites are most commonly associated with the forests flanking river systems in the Deep South. Mississippi Kites are also at home in riparian woodlands of the southern plains and locally in the desert Southwest (and more recently, golf courses whose shelterbelt landscaping mimics riparian habitats, and towns where mature trees abound). White-tailed Kite habitat is more xeric. Dry, open grassland, or weedy edges of agricultural land, all larded with a scattering of perches, are all the species asks.

But whatever and wherever the underlying terrain, the true habitat of the pointed-winged kites is open sky. Few (perhaps no) creatures on this planet are more ideally suited to this environment.

MIGRATION

Even though kites move great distances outside their normal range (as dispersing young in summer and early fall, or as overshoots in spring), kite migration is mostly confined to the southern states in which these birds breed, Mexico, and Central America. Large flocks of Mississippi Kites (some in excess of 1,000 birds) are seen in Texas, Mexico, and Panama in spring and fall (although in the United States, flocks of 50 to 300 are more common). Traveling in kettles, these Mississippi Kites pass Texas hawk-watch sites from mid-August through early October. In spring, the peak movement occurs more swiftly from late April through early May.

Individuals wander widely in the spring—probably more widely than is known since kites fly very high and, even if sighted, confusion with Peregrine Falcon is likely. More than a dozen individual Mississippi Kites have occurred in Cape May in a single spring, and the bird is a regular spring vagrant along the southern Great Lakes and Cape Cod, Massachusetts.

Swallow-tailed Kites are more geographically restricted in migration. Beginning in late July, many and perhaps most of North America's Swallow-tailed Kites gather in southern Florida, staging for migration. Roost totals of adults and young may approach 1,250 birds. In August, the birds depart, presumably crossing the Gulf of Mexico by unknown routes, and also presumably in flocks. Some, perhaps

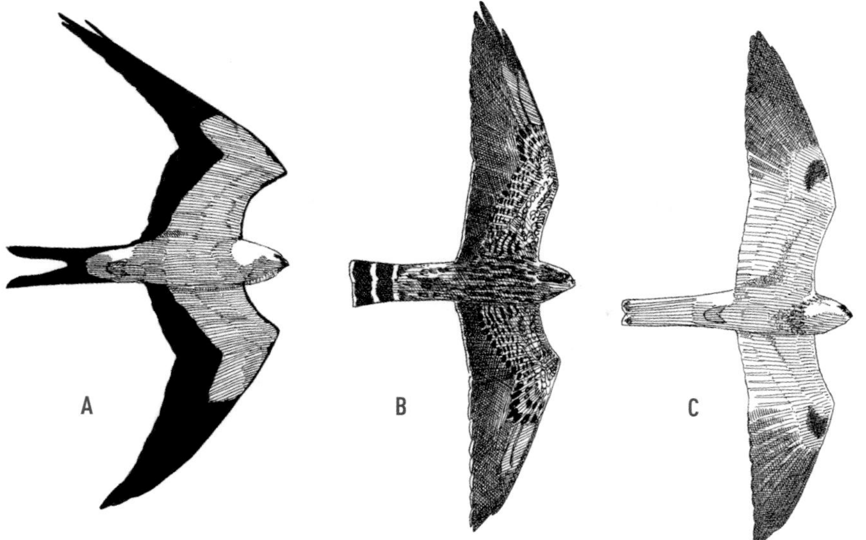

Kites: gliding juveniles of all species. American Swallow-tailed Kite **(A)**. Mississippi Kite **(B)**. White-tailed Kite **(C)**. These species are all distinctive; Swallow-tailed and White-tailed are difficult to confuse with any other bird. Mississippi Kite is easily confused with small and slight male Peregrine Falcon, but notice slightly flared tail with narrow pale bands, short outermost primary, different head pattern, unbarred primaries and secondaries, wings narrower at base than at wrist, and different flight style.

most, return by crossing the Gulf. Birds have been seen making landfall on the Dry Tortugas in April, and aggregations in excess of 200 birds have been seen in southern Florida (also in mid-April), but some Swallow-tailed Kites also use the land route through Mexico.

While restricted in its range, and mustering a North American breeding population of fewer than 1,500 pairs, the Swallow-tailed Kite is nonetheless a regular wanderer to points outside its normal range. Highly conspicuous and easily identified, the birds have been recorded throughout the Midwest and into northeastern Canada. In Cape May, New Jersey, they are annual, with two to five birds occurring most frequently between mid-April and mid-June.

White-tailed Kite is mostly sedentary, although it's nomadic in times of food shortage. Small numbers (50 to 75 birds per year) are annually recorded at the Golden Gate Hawk Watch in San Francisco each fall. In the 1970s, the bird's range was expanding north along the Pacific Coast as well as east along the Gulf Coast and into Florida (the species' historic range). This Florida expansion appears tenuous. Vagrants have occurred as far north as Minnesota, New Jersey, Connecticut, and Massachusetts (and there is one breeding record for North Dakota). In spring and fall, this species should be watched for almost anywhere.

IDENTIFICATION

All three pointed-winged kites are easily separated from one another: Swallow-tailed Kite by size and shape, and Mississippi Kite and White-tailed Kite by plumage, behavior, and, over much of North America, distribution. The challenge in

An adult White-tailed Kite. The pointed-wing kites are all studies in black, white, and gray. Few would argue against kites being the most graceful and elegant of raptors. NH

kite identification lies in distinguishing distant kites from *other* birds (including some species that are not even raptors).

THE GENERIC POINTED-WINGED KITE

This kite type is a slim, finely proportioned, medium to medium-large raptor with long falconlike wings and a long tail that twists and turns in flight. Its primary colors are a study in white, black, and gray.

Flight is light, buoyant, nimble, acrobatic, and fluid with wingbeats that are slow, stiff, and deliberate.

Mississippi Kite

This falcon-shaped kite has a wide but interrupted range across the southern United States. It can be conspicuous in the shelterbelt habitats of Texas, Oklahoma, and southeastern Colorado, and in the riparian woodlands that flank desert watercourses in Arizona and New Mexico. Although numerous in the swamp woodlands of Georgia and South Carolina, it is often very *inconspicuous,* in part because kites spend most of their time high above the canopy and human observers do not.

Perched, most often in a mature deciduous tree, the bird can appear deceptively small — hardly larger than American Kestrel. In flight, the bird's long wings and fanned tail put it in a size class with Broad-winged Hawk. The birds breed and hunt communally. Hunting aloft, plucking insect prey out of thin air, birds are most active from midmorning to midafternoon, when thermal activity peaks. During migration, Mississippi Kites form large flocks that pass through southern Texas in late August through early October. In spring, the birds return in flocks in April and May.

More than the much larger Swallow-tailed Kite and the ground-focusing White-tailed Kite, Mississippi Kites specialize in the capture of large flying insects, primarily dragonflies, beetles, and cicadas. Mississippi Kite's diet also includes assorted reptiles, amphibians, small birds, and mammals, but insects are its staple, and these the bird consumes in the air.

IDENTIFICATION. The Mississippi Kite is a medium-sized falcon-shaped bird with the size, proportions, and shape of a Peregrine. Adults in bright sun are medium gray above, light gray below, and conspicuously white headed. The tail is uniformly blackish above and below. From below, dark flight feathers contrast with the gray body and wing linings. Above, pale secondaries form a prominent white patch on the trailing edge of the wing (particularly prominent on males) — given the bird's acrobatic flight, this mark is far more conspicuous than many field guides allow.

Unless backlit or very high, confusion with Peregrine Falcon is not likely. Juvenile Mississippi Kite is a different story — brownish gray above, charcoal and chocolate below. The belly and breast are lightly to heavily streaked cinnamon brown. The tail is dark with narrow, translucent white bands *that are very obvious, particularly when birds are backlit.* The underwings are very patterned but usually appear all dark, almost black.

Subadults (second-year birds) have a head and body that are gray like adults', but wings and tail of juveniles. Some subadults seen in May or June may still retain some streaking on the underparts. The pale head and white wing patches of adults may not yet be prominent, but the narrow white bands on the tail remain very conspicuous.

Insofar as most spring vagrant Mississippi Kites are subadults, this transitional plumage is one to commit to memory.

IN FLIGHT. The body of Mississippi Kite is long, narrow, and slight. In flight, the wings are long, slim, tapered to a sharp point, *but oddly proportioned.* The arms are long, but surprisingly narrow. The equally long hands are uncharacteristically broad.

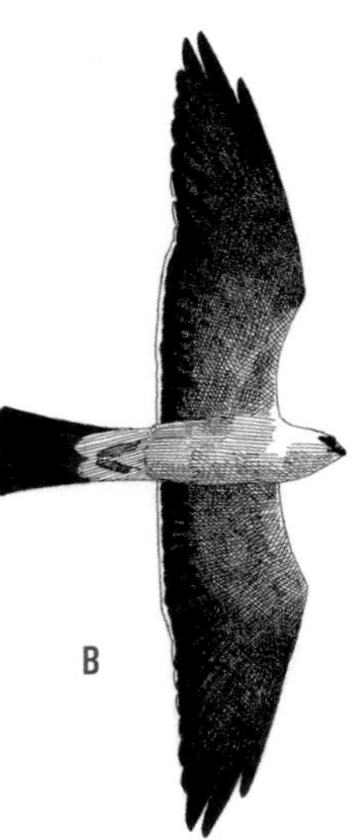

Mississippi Kite, underside. Juvenile **(A)**. Adult **(B)**. Slim, long winged, medium-sized, with slender body and fairly long tail. Wings are narrow armed and broad handed, with distinctly short outermost primary. Tail flares slightly at tip and is constantly spread and twisted in flight. Adult is gray bodied with darker underwings, black tail, and whitish head. Juvenile has body and underwing coverts buffy, with dark streaking and spotting of variable density; may be lighter or darker than illustrated; tail has several narrow light bands; flight feathers dark gray, with pale bases on outermost primaries; dark cheek patch sets off pale eye stripe. Distinguished from juvenile Peregrine by shape and flight style, narrow pale bands in tail, head pattern, and pale bases on outermost primaries.

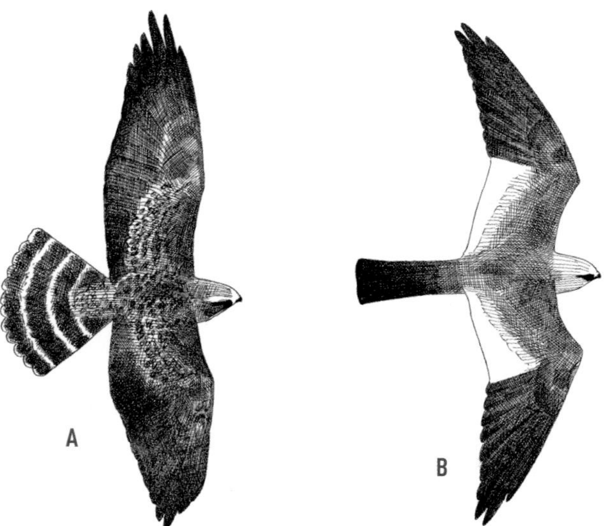

Mississippi Kite, upperside. Juvenile **(A)**. Adult **(B)**. Bright silvery secondaries of adult are distinctive. Juvenile has wing coverts and all feathers of back edged with buff. Palest juvenile Peregrines can also have scaly buff markings above, but more uniform.

Mississippi Kite going away, with long wings characteristically pressed down and tail tilted for control.

This gives soaring birds the appearance of having overdeveloped forearms — an avian "Popeye the Sailor" look. With wings turned back, gliding birds appear falconlike.

The leading edge of the wing bulges forward at the wrist. The trailing edge is straight-cut when the bird is executing a shallow glide (which it usually is). In shape, the wing is similar to that of a Broad-winged Hawk in a glide, but it is both narrower and longer. The outermost primary is conspicuously shorter — it looks like it is just growing in. *It is often splayed from the wing and very conspicuous, even at considerable distances on soaring birds.* By comparison, the outer primaries of Peregrines are all of near equal length and never splayed. The outermost primary on Peregrines is difficult to distinguish.

A Mississippi Kite's tail is long and flares outward toward the square-cut tip. Particularly apparent when the birds are soaring, the flared outer tail feathers are apparent even when the tail is closed. Among North American raptors, the shape is unique but not above confusion.

The wingbeat of a Mississippi Kite is slow, stiff, deep, and deliberate, with little movement or flexibility down the length of the wing. The bird gives the impression that flapping is something it does grudgingly (as if exerting effort is beneath its acrobatic dignity). When the bird does resort to powered flight, it accelerates quickly or uses this application of force to maneuver deftly.

Mississippi Kite controls momentum by manipulating the length of its wing; a twisting and turning tail that seems to micromanage the air aids with tight maneuvers. This yawing tail motion is very distinctive and nearly idiosyncratic (although Swallow-tailed Kite does it as well).

Soaring must seem like cheating to Mississippi Kites. They seem loath to do it, much preferring to sail, glide, and swoop, which they will do for hours on end.

1. An adult Mississippi Kite, head-on, with wings slightly down-curved, tail tip modestly flared, outermost flight feather lifted and separated (like a lifted pinkie). Arizona, June NH

2. A soaring adult Mississippi Kite with white head, gray underparts, and all-black tail. The shape is very falconlike, but take a close look at the wing shape, in particular the short, narrow arms and long, broader hands. Other key points include the shorter outer primary and the slightly flared tail. Arizona, June NH

3. An adult Mississippi Kite caught in a downstroke. Wingbeats are deep, stiff, and slow. The whitish head and (especially) the white secondaries are evident at a distance. Not so the touch of bay in the primaries. Arizona, July NH

4. An adult Mississippi Kite in a glide. Wings are angled slightly down when gliding or "sailing," and to a lesser degree when soaring. Arizona, June NH

5. A one-year-old Mississippi Kite showing juvenile wing and tail pattern but (for the most part) the head and body plumage of an adult. The shape remains the same. Cape May, June DF

6. A one-year-old Mississippi Kite in a characteristic fast glide, frequently seen as they hunt dragonflies. Except for the flared tail, it appears very falconlike. Cape May, June DF

7. A banking one-year-old Mississippi Kite. Wing shape can recall Broad-winged Hawk at times, which can also be seen in #2. Cape May, June MOB

8. A one-year-old Mississippi Kite in heavy molt but still mostly in juvenile plumage (note the streaking on the breast). The normally clean wing shape is altered by molt. Cape May, June MOB

9. The same individual as #5, soaring and showing a year's worth of wear on those flight feathers. Notice the characteristic translucent tail bands. DF

Swallow-tailed Kite

Few would disagree that the Swallow-tailed Kite is the most graceful flier of any North American raptor. This distinctive aerial predator will glide and float over the canopy on wings that cancel gravity and suspend time, riding the lightest breeze to snatch dragonflies from the air or pluck prey from treetops and consume it on the wing. Less commonly does the bird spread sail and soar. Even more rarely does it flap.

Some might argue that Swallow-tailed Kite is also the continent's most beautiful bird. Elegant, almost rakish in design, it dresses formally in black-and-white attire (tails and all). In good light, the bird's dark upperparts appear bluish gray and hint of iridescent hues.

Much larger birds than Mississippi Kites, Swallow-tailed Kites seem to fly in slow motion with movements that are similar, but the flight has a dreamlike quality that makes the flight of Mississippi seem jerky and erratic. Swallow-taileds also commonly hunt lower than Mississippi Kites — often just above the treetops — where they pluck lizards, snakes, and birds from among the leaves.

When Lewis and Clark went exploring, they discovered the Swallow-tailed Kite north to Minnesota and west into Kansas and Nebraska. Now, this species' North American breeding range is limited to the coastal plain from South Carolina to Louisiana (with a few pairs in Texas and a recent range expansion into southern North Carolina), but Florida is the bird's stronghold. Fully two-thirds of the United States breeding population is found there.

On territory, the birds are rarely alone. But beginning in August, the entire population is extremely social and vacates North America en masse, most flying across the Gulf of Mexico to winter in South America. A few individuals follow the land route through Texas, Mexico, and Central America, and some return this way.

Swallow-taileds are early spring migrants, arriving as early as late February and early March in southern Florida and continuing into May. Overshoots, appearing in states and provinces well north of the birds' breeding range, are most often recorded between April and early July. Staging for fall migration begins in late July. Most birds leave the United States in August and all are gone by mid-September.

IDENTIFICATION. This is a large and distinctive raptor — almost as large as an Osprey (one of the very few birds it could conceivably be confused with). Males and females are alike in plumage and size. Juveniles have pale feather edges on the back, and black areas are muted with brown. Also, the forked tail of juveniles is consider-

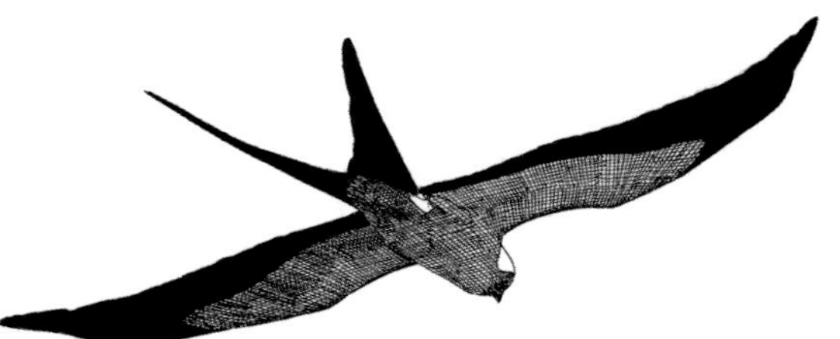

Swallow-tailed Kite going away.

Swallow-tailed Kite, underside. Nearly unmistakable, with bold black-and-white pattern, very long wings, and long forked tail.

Swallow-tailed Kite, upperside. Glossy blue-black above with darker shoulders and white head.

Swallow-tailed Kite. Incredibly graceful and elegant in flight.

ably shorter than the tail of adults, but still conspicuous and distinctive. In a glide, the bird holds itself aloft on down-bowed wings — reminiscent of Osprey and Magnificent Frigatebird (another large fork-tailed bird that Swallow-tailed Kite might be confused with).

But, seen clearly in flight, at anything approaching a reasonable distance, a Swallow-tailed Kite should be instantly identifiable. There seems no reason to belabor this. Other birds might have characteristics that hark to Swallow-tailed Kite, but nothing else really looks like Swallow-tailed Kite. Distance is, of course, the giant equalizer, and enough distance will cause even something as distinctive as a Swallow-tailed Kite to be overlooked (maybe even misidentified).

Hint: Check all distant Osprey. Head-on, the size and figuration of Osprey and Swallow-tailed Kite can be disarmingly similar.

1. A Swallow-tailed Kite in a full soar. Any questions? Florida, April CS

2. A Swallow-tailed Kite in a shallow glide, showing summer wing molt. South Carolina, July CS

3. A Swallow-tailed Kite gliding. In cumulus-filled southern summer skies, high birds in shadow can appear as all-dark silhouettes, despite the white body and wing coverts. Swallow-tailed Kites soar very high and glide fast. Although sometimes overlooked because of their altitude, they are very nearly unmistakable. Florida, April CS

4. The dorsal view of a Swallow-tailed Kite banking in a soar. The birds soar on flat wings with wingtips very gently raised. Florida, May NH

5. A Swallow-tailed Kite in a shallow glide, or "sailing." When gliding, wings are down-turned, after the fashion of Osprey — in fact, distant birds can be overlooked, mistakenly dismissed as Ospreys. Florida, May NH

White-tailed Kite

There is something perfunctory, even businesslike about White-tailed Kite. Mississippi and Swallow-tailed kites exude a sense of casual ease in the air. The procurement of prey seems almost like an afterthought.

Not so the energetic White-tailed Kite. When it hunts, it focuses. A hundred feet up. Head turned down. Eyes fused to the grass below. Wings turned into the wind, fluttering like banners. If the immediate patch doesn't produce, the bird moves on, flying on kite-stiff wingbeats to the next promising spot (usually a short distance away). If the bird sees prey, it parachutes closer in a controlled, vertical glide, and then . . . stops. If the distance is still too great, it parachutes again . . . and stops.

When the bird is about ten feet above the ground, and success seems assured, the White-tailed Kite lifts its wings high over its body, extends its feet, and collapses (as much as plunges) atop its food.

More than half the time, the kite's calculated stoop is successful. Almost always, the prey is a small rodent, although White-tailed Kites do occasionally take other prey (including some reptiles and insects). Open country — pastures, grasslands, airports, overgrown agricultural edges, and the grassy edges of highways — is the bird's preferred habitat.

A century ago, the White-tailed Kite was considered rare in North America (some thought doomed). Once found from the Carolinas to California, habitat alteration and human persecution reduced its numbers until, by the early twentieth century, the White-tailed Kite was found only in California and southeastern Texas.

Just prior to midcentury, the population began rebounding in California, followed by a range expansion into Oregon and southeastern Washington. Simultaneously, small numbers of birds began recolonizing former breeding areas in Florida (although this trend has recently slowed or reversed). Today, as nesting and wintering species, White-tailed Kites are found in California and Oregon and across much of the South into Mexico and South America (where their numbers are also increasing). As vagrants, the birds have appeared irregularly but widely.

Unlike the other two pointed-winged kites, White-tailed Kite is nonmigratory. The birds are somewhat nomadic, moving to locations that offer a surfeit of prey during and/or after the breeding season. Birds regularly winter in areas where they do not (or do not yet) breed; in fall, small numbers (fewer than 100) White-tailed Kites are seen passing Golden Gate Raptor Observatory counters in California and a few (fewer than 10) are regularly recorded at Smith Point and Hazel Bazemore, Texas.

In winter, the birds roost communally (in groups of 100 or more). Crepuscular to a degree not seen in other kites, White-tailed Kites hunt actively at dawn and dusk. They commonly spend midday on a perch — a time when Mississippi and Swallow-tailed kites are hunting aloft.

IDENTIFICATION. The White-tailed Kite is a medium-sized raptor, about the size and shape of a Ring-billed Gull (and the similarities do not end there). For a time, the bird was considered conspecific with the stockier and shorter-tailed Black-

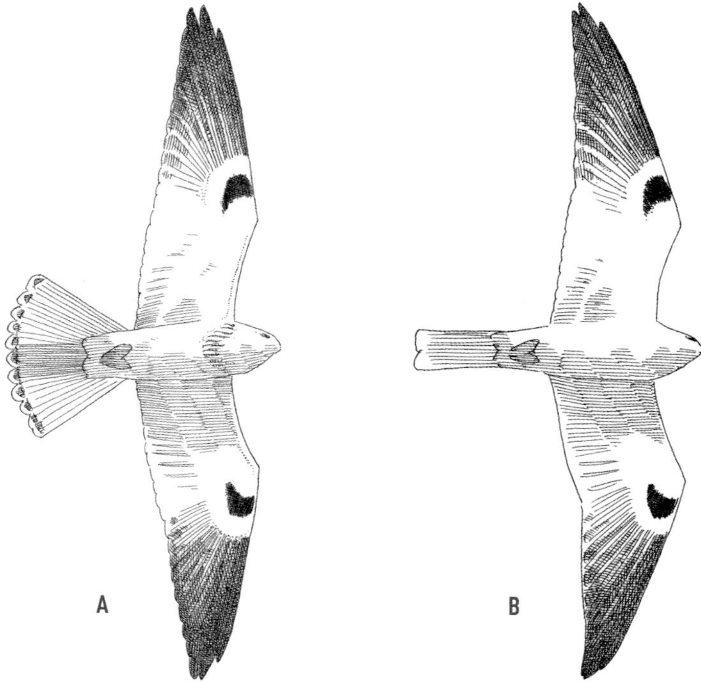

White-tailed Kite, underside. Juvenile **(A)**. Adult **(B)**. Slim and long winged, with very narrow tail; wings never quite straight, even when soaring, as in **(A)**. More likely to be mistaken for a gull than for another raptor. Juveniles differ only slightly from adults, with buffy streaked breast band and dusky spot at the tip of each tail feather.

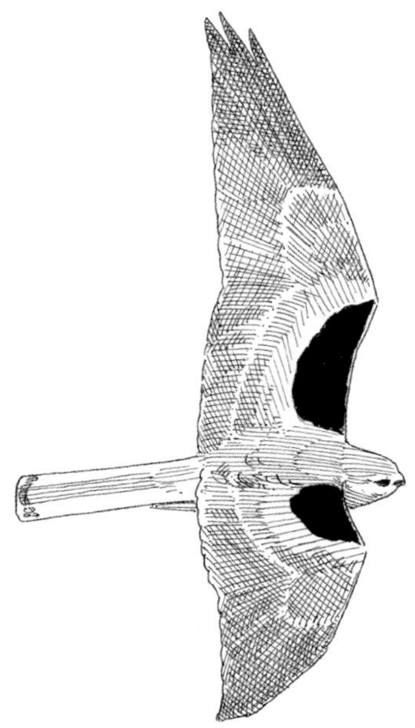

White-tailed Kite, upperside. Black shoulder patches and whitish tail distinctive in all plumages. Juvenile (shown) has dusky tail tip, buffy wash on crown, brown spotting on back, and buffy tips on wing coverts. Adults are uniformly pale gray, lacking all these markings. In known pairs females can be identified by slightly grayer shoulder patches.

Juvenile White-tailed Kite, hovering.

shouldered Kite of Australia, southern Eurasia, and Africa, but reevaluation in 1992 returned the bird to its original separate species status.

The bird is strikingly and unmistakably patterned. Adults are pale gray above, with a white head, white tail, and white underparts. Above, bold black patches dominate the upperwing coverts. Below, charcoal wingtips are punctuated by a bold black comma at the wrist.

Juveniles have a ruddy-tinged necklace on the breast, a brownish scalloped crown and back, and a dusky tip to the tail.

IN FLIGHT. Like Mississippi Kite, White-tailed Kite is falconlike, but overall more robust or more front-heavy — large headed, big handed, and with a tail that seems too narrow for the rest of the bird. It does everything with a dihedral. When gliding (or soaring, kiting, or hovering), its falconlike wings are usually held in an exaggerated U-shaped dihedral — hunched at the shoulders, less acutely uplifted along their length, then flared upward at the tip. As this species glides, the wings are acutely pointed; while it is hovering, the wings appear more rounded or blunt.

When the bird is gliding, the closed tail is very narrow, and tapers toward the notched tip. It is proportionately (and distinctly) shorter than the tail of an adult male Northern Harrier, a bird with which it is frequently confused, owing to nominal similarities in plumage and both birds' penchant for low, cruising flights over open grasslands.

Flight is light, buoyant, and direct. Wingbeats are slow, shallow, and stiff, somewhat gull-like but, in a glide, the pronounced dihedral and tendency to rock unsteadily impart a nighthawklike GISS (pronounced "jizz"): general impression, size, and shape.

The bird hovers frequently and energetically, recalling an oversized American Kestrel. The body is sharply angled — head up, tail down. The head remains stiff and fixed. To study something below, the bird repositions its body, not its head. The wings flick down and forward with all the energy concentrated in the arms. The broad hands merely flick — they seem to be going along for the ride.

Gliding birds appear mostly light with black areas restricted to the shoulders and wingtips. Distant, hovering birds show more extensive black.

When holding into the wind (kiting), the secondaries ripple or flutter.

1. An adult White-tailed Kite in direct flight. A beautiful and distinctive bird. When seen this well, virtually unmistakable. Texas, December NH

2. An adult White-tailed Kite. This buoyant and graceful species glides and sails far more than it soars. Texas, December NH

3. The dorsal view of an adult White-tailed Kite in a fast glide, showing the two plumage traits that not only add to the bird's character, but also give it its names. Called White-tailed Kite now, the bird was formerly known as Black-shouldered Kite. Note how acute the dihedral is on this individual — Turkey Vulture–acute. More typically, the wings are held in a distinctive U-shaped configuration. White-tailed Kite does everything (except perch) with its wings elevated above the body. Texas, February TJ

4. A juvenile White-tailed Kite banking as it leaves the hover. A well-named species. Note the remnant pale, beaded necklace, which is worn by older juveniles. California, April NH

5. An adult White-tailed Kite hovering, a common and characteristic hunting behavior. White-tailed Kite is the only kite to hover on high. It can recall American Kestrel, but the flap in the hover is slower and looser. California, October NH

6. A juvenile White-tailed Kite showing classic hovering profile and classic rusty (to buffy on some individuals) necklace pattern that birds carry for only a short time. Note, too, the dark beaded tail feather tips. In short order, the bird will look like #1. California, May NH

7. An adult White-tailed Kite gliding, showing diagnostic dihedral. Among all raptors, the elegant White-tailed Kite might only be confused with an adult male Northern Harrier. Texas, December NH

PUTTING IT TOGETHER: Telling Kites From Other Raptors

While each of the three pointed-winged kites is easily distinguished from the others, all have traits that might cause them to be confused with or mistaken for other species. All three birds have characteristics that are shared by gulls, for example. Given the right circumstances, Mississippi Kite may be mistaken for not only Peregrine but even the larger Swainson's Hawk, a buteo.

Mississippi Kite Versus Peregrine Falcon and Swainson's Hawk

Separation of Mississippi Kite and Peregrine Falcon is a raptor identification classic.

The average Peregrine is considerably larger than a Mississippi Kite. Smaller male Peregrines approach Mississippi Kites in size and slim proportions, but Peregrines will always be more heavily proportioned with broader wings and a heavier, chestier body.

Adult Mississippi Kites are always readily distinguishable from Peregrines by plumage, but juvenile and subadult kites and juvenile Peregrines share more plumage similarities. While first-year Mississippi Kites are somewhat variable, ranging from sandy tan (with darker streaking) to chocolate brown all over with little streaking, it is these dark juvenile Mississippi Kites that are most easily confused with Peregrines.

When distance is great enough to obscure the face pattern of a juvenile Peregrine, a juvenile kite and a juvenile Peregrine will appear much the same color. At closer range, the transparent white barring on a juvenile kite's tail is diagnostic. A Peregrine's tail may be barred, but the dark markings appear on a brown backdrop. They are not translucent, and they do not stand out.

The wings of the birds are shaped differently — kite narrow at the arm, wide at the hand; Peregrine wide at the arm, narrow at the hand. This difference should stand out but, for some reason, it often does not. What does stand out, particularly when gliding, is the kite's single shortened anterior flight feather.

Seen at a reasonable distance, the anterior flight feather on soaring birds juts out, like a feather about to be molted (or one just coming in). In a full soar, the outer flight feathers, too, are slotted and fingerlike. The outer primaries on Peregrines appear seamlessly fused.

In a glide, the shorter outer primary on the kite makes the wing look unfinished and incomplete. There's a gap in the leading edge of the wing, an obvious piece missing near the tip. The leading edge of a Peregrine's wing is, again, seamless.

Mississippi Kites and Peregrines also differ in the shape of the tail. The kite's tail is proportionately longer and splays out toward the tip — a dead giveaway. The tail of a Peregrine remains straight along its length, if it doesn't taper toward the tip.

A Mississippi Kite's tail is also more squared off. The tail of a Peregrine is rounded and commonly notched when folded. Juvenile Peregrines also have a broad, buffy terminal band, a feature that is lacking on kites. Finally, the tail of a Mississippi Kite is almost always in motion, twisting and turning like a long-bladed leaf in the wind. A Peregrine's tail is, by comparison, rigid.

Behavior probably provides the best clue to the identity of a distant falcon or kite. Most apparent is the difference in wingbeats. The wingbeat of a Mississippi Kite is stiff, deep, languid, and somewhat hesitant or halting. The wingbeat of a Peregrine is steady, powerful, rapid, fluid, and shallower — the entire wing ripples with

Adult Mississippi Kite catching a dragonfly in midair, a typical behavior for this species, only rarely shown by Peregrine Falcon.

each stroke, and each stroke seems to build upon the last, gaining power, energy, and speed. There is nothing hesitant about it.

Peregrines flap much more than Mississippi Kites. Kites move their wings as if it were an afterthought or a last resort. In addition, kites prefer to glide rather than truly soar. Peregrines soar frequently.

It is fortunate that this kite tends not to soar, because it is precisely when Mississippi Kite does soar that it looks most like Peregrine. If the bird is overhead, the differences in the lengths and spacing of the outermost flight feathers should be apparent. The translucent white bars on the juvenile's tail will also be distinctive.

When a soaring Peregrine is viewed wing-on, the wings are flat (sometimes upswept at the tip), whereas the Mississippi Kite's wings will look somewhat droopy or down-curved.

Wing position is also helpful when the bird is gliding or sailing. Peregrines show an abrupt upward tilt at midwing, at the joint. The tilt is not static. It lifts and straightens as the bird maneuvers, almost like the flaps on an airplane. The wings of Mississippi Kite curve down gently and evenly throughout their length.

The proportional and plumage similarities between Mississippi Kite and juvenile Swainson's Hawk have been noted. What's more, both are flocking species, and both migrate at much the same time.

Happily, the two species are markedly different in size, with Swainson's Hawk being larger. Confusion between the two species most commonly occurs in the case of individual birds seen soaring at great distances or sighted soaring directly overhead.

At a distance, the different sets of the wings should be readily apparent. Swainson's Hawks hold their wings uplifted in a V. With Mississippi Kite, it's just the opposite: kite wings droop.

Directly overhead, the wings of both Mississippi Kites and Swainson's Hawks are long and symmetrically tapered. Both have short anterior primaries. In addition, juveniles of both species are heavily streaked below. Both have a dark trailing edge to the underwing. In sum, the birds have several common traits.

From below, the characteristic that most easily distinguishes juvenile Mississippi Kite and Swainson's Hawk is the tail. Both tails may appear overall dark, but the narrow, translucent light bands on the tail of juvenile Mississippi Kite will be particularly prominent at this angle. Also, and at great distances, the profile of a

Swallow-tailed Kite may be most easily confused with Osprey, although the resemblance is not very strong.

soaring Anhinga and Mississippi Kite can be surprisingly similar. But an Anhinga would have to be very far away for the projecting neck to be invisible.

Swallow-tailed Kite Versus Osprey and Magnificent Frigatebird

Among raptors, the Swallow-tailed Kite can only be confused with a distant Osprey. Osprey and Swallow-tailed Kite both have dark upperparts, white underparts, and swept-back, down-curved wings.

But the Swallow-tailed Kite has a curve throughout the length of its wing, both in ventral profile and head-on, and never shows the angular, M-shaped configuration of an Osprey. The dissimilarities between Osprey and Swallow-tailed Kite far outweigh any commonality. The best way to avoid confusion is simple awareness and something more than a quick, dismissive glance. Sometimes it's not "just another Osprey."

Among nonraptors, Magnificent Frigatebird, another fork-tailed aerial master, bears more than a superficial likeness to Swallow-tailed Kite (particularly juvenile frigatebirds, seen head-on, which show white heads and partially white underparts). Again, mindfulness is probably the best defense. Frigatebirds are *considerably* larger than Swallow-tailed Kites and have a very prominent, projecting head. In addition, and except during migration, Swallow-tailed Kites are unlikely to be seen offshore. Frigatebirds, on the other hand, while much at home over open water, are frequently seen well inland.

White-tailed Kite Versus the Field

White-tailed Kite shares several characteristics with other North American raptors but is too dissimilar to American Kestrel and adult male Northern Harrier to be truly confused with them. *Rule of thumb: When in White-tailed Kite country, not all*

Adult White-tailed Kite going away.

birds that hover are automatically kestrels, and not all gray-and-white birds that fly with up-angled wings are male harriers.

And, obvious or not, Northern Harrier doesn't have black shoulders; White-tailed Kite doesn't have a contrasting white rump patch.

White-tailed Kite most closely resembles the adult Ring-billed Gull (*Larus delawarensis*) and Mew Gull (*L. canus*). The two gull species are roughly the same size as the kite and are gray above and white below, with long tapered wings that are dark at the tips. Caspian Tern, stern-on, might also be a candidate for confusion.

But gulls and Caspian Tern have very large heads that are roughly as long as their tails. Next to gulls, kites are comparatively small-headed birds that have very long tails. Gulls also have a downward crook in the wings. White-tailed Kites fly with a dihedral. And although gulls can and do hover, they do so only for short durations and with obvious effort. Hovering is as natural as flight to a White-tailed Kite, and flight is as natural to a kite as movement is to the wind.

Northern Harrier

The Great Fooler

SPECIES
Circus cyaneus

Top to bottom: juvenile Broad-winged Hawk, juvenile Coopers Hawk, adult male Northern Harrier; all soaring.

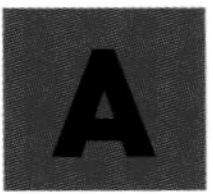

A slim, graceful raptor of open areas, the Northern Harrier is North America's sole representative of a genus whose 16 species claim territory on every continent except Antarctica. Perhaps because it is common throughout much of its North American range and conspicuous in open haunts (and perhaps, too, because its characteristic low, cruising flight appears casual and undashing, compared to other raptors), the bird is underrated. Its lack of flamboyance notwithstanding, the harrier is nevertheless an admirable, versatile, dogged, and enigmatic raptor. If you were to study the many facets of raptordom — from migration to breeding strategy, venatic versatility, and the like — and trace the lines of evolutionary development to their culmination, your search would probably end with the Northern Harrier.

As a hunter, the harrier is a master of the subtle art of sneaking up from behind; its standard mode of hunting seems to echo the old adage, "Nothing ventured, nothing gained." A low, cruising flight is punctuated at intervals by pull-ups, hovers, wing-overs, and drop pounces — just the strategy needed to trap a tide-flushed meadow vole or a luckless young Red-winged Blackbird.

Many other birds of prey approach the business of hunting by perching on the sidelines, waiting for opportunity. Harriers hunt.

LEFT: An adult male Northern Harrier from above. The white rump may be the single best classic field mark shown by any raptor. The amount of brown in this individual's plumage, as opposed to full gray, suggests that it is in its first adult plumage. JL

As a breeder, a wintering species, and a migrant, Northern Harrier is found throughout North and Central America. Only the High Arctic is deprived of the bird's presence. Its breeding range is extensive — from Alaska to Newfoundland in the north, Virginia to Texas to California in the south. Except for coastal British Columbia and the Maritime Provinces, the bird largely vacates Canada in winter. It also shuns the cold and snow of the northern plains states.

Harriers winter from the Maritime Provinces of Canada to the Great Lakes, and — in a line that seems to skirt those states where snow on the ground lasts till spring — west to the Pacific. Some harriers migrate as far south as Colombia, but most seem to find a winter territory somewhere short of South America, apportioning themselves over much of North America wherever they can find suitable habitat: prairies, open agricultural land, freshwater and saltwater marshes, sagebrush, and semiarid grasslands, any place that promises a surfeit of prey and gives a bird lots of unencumbered space to stretch its wings.

In winter, Northern Harriers often roost communally on the ground. Concentrating in areas where rodent populations are high, roosts of more than 1,000 birds have been recorded, but 10 to 20 birds are more common. Larger females and juveniles tend to hunt in more open areas. Small, more agile adult males commonly hunt in tighter confines — upland edges and smaller, weedy fields — and may prey upon birds more than females do.

In spring, the birds go ballistic. Males and females on territory engage in dramatic aerial displays. Called *sky dancing,* these courtship and territorial displays involve a series of undulating climbs and dives, capped with backflips at the peak of each ascent.

And on top of all this, and unlike some other more celebrated birds of prey, harriers seem also to be gifted with an uncommon measure of intelligence. Experienced raptor banders will tell you that capturing Peregrines and Red-taileds is no great feat. Catching an adult male Northern Harrier takes more than luck.

MIGRATION

Spring and fall migration periods are protracted. In fact, the Northern Harrier may have the longest overall migration period of any North American raptor. In late February and early March, adult males begin to pass through coastal and interior sites. Females follow in mid-March through April. Subadults continue to wander north through May and June.

Failed breeders and nonbreeders can be seen passing hawk watches as early as mid-July. Most juveniles migrate from August through September. Adult females commonly migrate in October, and adult males are most frequently seen from late October through December.

In migration, the Northern Harrier is often the first bird of the day recorded at a hawk watch and just as often the last. Not surprisingly, they are crepuscular and share several convergent evolutionary traits with owls (such as a sound-trapping facial disk and an acute sense of hearing). Harriers have been seen coming in from the ocean before sunrise, hunting on moonlit nights, and twice migrating through the beam of the Cape May Point lighthouse in full darkness.

Harriers commonly migrate in weather conditions that discourage other raptors, such as fog, snow, sleet, rain, and even full-blown gales. They seem undaunted

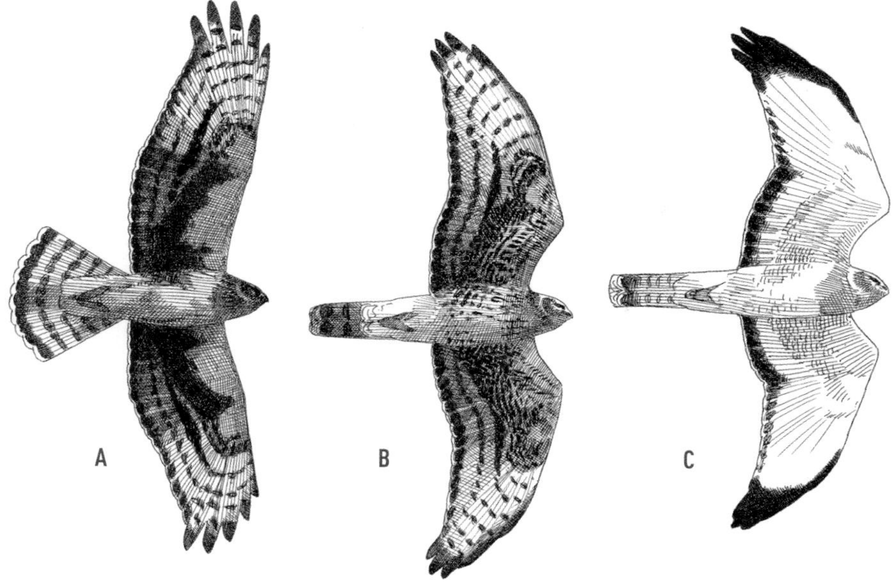

Northern Harrier, underside. Juvenile **(A)**. Adult female. **(B)** Adult male **(C)**. Shape always distinctive: long primaries; long tail; slender body; long narrow wings; tail often held closed when soaring, and even then the bird rises faster in thermals than other raptors. Females are noticeably larger and bulkier than males. Adult males are striking, with clean white body and wing linings, gray breast, and black tips on secondaries and outer primaries. Adult females are pale buffy, with heavy spotting and streaking on the body and underwing coverts, jagged barring on flight feathers, and dark secondaries. Juvenile is like female but has unstreaked orange on body and underwing coverts; very dark secondaries and greater coverts create a dark patch on inner wing more prominent than on female.

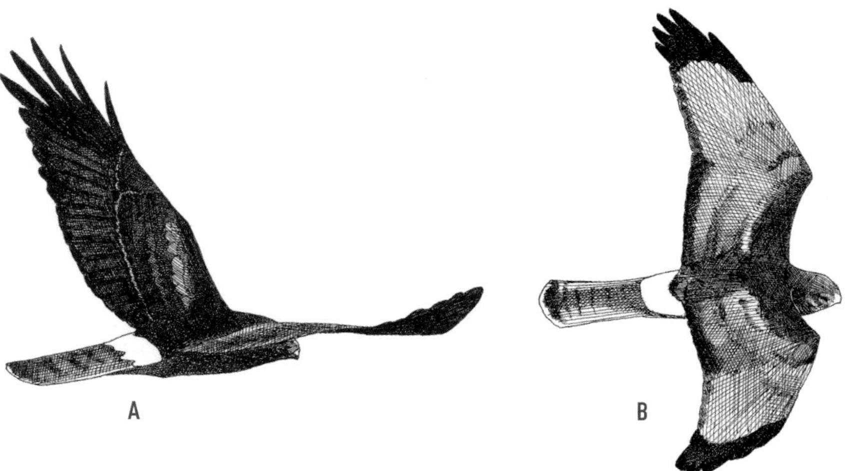

Northern Harrier, upperside. Juvenile **(A)**. Adult male **(B)**. White rump is always conspicuous. Adult male is silvery gray with variable brown back and lesser coverts and with black tips on secondaries and outer primaries, as on underwing. Juvenile is dark brown with buffy tips on greater coverts and pale buffy bar across lesser coverts. Female is similar but paler and more spotted; flight feathers grayer and barred.

by over-water crossings up to and exceeding 40 miles. Despite the harrier's cavalier attitude toward open water, hawk-watch sites associated with water barriers do, nevertheless, commonly record greater numbers of harriers than ridge sites, possibly because coastal areas often offer harriers prime, open hunting habitat. While harriers do make use of light to moderate updrafts, their light weight and very low wing-loading make them unsuited for flight in strong turbulence.

On the interior ridges, harriers often ignore updrafts and fly wide of the ridge or high overhead, choosing to use thermals or even their own powered flight for lift. More than all raptors save falcons and, to a lesser degree, Rough-legged Hawk, Northern Harrier seems to be most at ease flying under its own steam.

Though harriers are commonly seen migrating alone, pairs and small groups of three to eight birds may be seen at prime hawk-watching locations. These groups usually travel single file. If you spot a migrating harrier, it is good practice to check ahead for a lead bird, then behind to see if your bird has a shadow.

IDENTIFICATION

The Northern Harrier is a lean, lanky, *long-winged, long-tailed,* medium-sized raptor. If you want to shortcut the entire identification process, simply study a harrier silhouette and fix it in your mind. While harriers have a number of traits that recall other species, taken in sum, *there is nothing else in North America that has the exceedingly long-winged, long-tailed profile of a Northern Harrier.* It is unique.

Adult plumages are sexually dimorphic. Males are silver gray above and, except for the gray head, white below — white body, white underwings (with black tips and a black trailing edge), and a grayish white tail. In hawk-watching circles, this distinctive plumage has earned the adult male Northern Harrier the nickname "the gray ghost."

Females are brown above, tawny below, with brown streaking extending down the sides of the chest and flanks. Females are noticeably larger and have wings more broadly proportioned than males.

Juveniles of both sexes are dark rufous brown above and cinnamon-orange on the wing linings and body below (getting paler toward the belly). The tail and the trailing edge of the wings appear dark.

Note: While juvenile harrier has a streaked breast and flanks, against the bird's cinnamon underparts the streaks become indistinct at any appreciable distance, so distant juveniles appear unstreaked and uniformly cinnamon below.

In spring, second-year (i.e., almost one year old) birds are paler because of feather wear. They may, in fact, suggest adult male birds. Most, however, will still show some warm or buffy hues — vestigial cinnamon.

In all plumages, there is a prominent, well-defined, fist-shaped white patch smack on the rump — a rump patch. Many other species of raptors show some white at the base of a dark tail, including Snail Kite, Rough-legged Hawk, Golden Eagle, Ferruginous Hawk, Swainson's Hawk, Red-tailed Hawk, Harris's Hawk, and juvenile White-tailed Hawk, but only the harrier has a clearly defined rump patch. *Caution: Cooper's Hawk, whose size (among females) and relative proportions approach Northern Harrier's, often shows bright white undertail coverts, which observers sometimes mistakenly project onto the bird's rump. Confusion is particularly likely when Cooper's Hawks are displaying (something they do at all times of the year) and fluff their undertail coverts expressly to be seen.*

IN FLIGHT. The wings of harriers are *exceedingly* long and proportionately narrow — very planklike when the birds are soaring, pointed and falconlike when the birds are gliding.

The harrier's tail is likewise long and narrow, very accipiter-like. Unlike most other raptors, this species frequently soars with its tail *closed,* but even fully fanned (which for harriers isn't very wide) the tail still appears long.

An adult male Northern Harrier flapping.

When the harrier is soaring, gliding, or hunting in level flight, it holds its wings in an acute uplifted V or dihedral, a field mark evident at great distances. Other hawks have a dihedral, but harriers lift their wings in a more exaggerated V than any other species except Turkey Vulture.

Harriers are very light birds — both in real weight and on the wing. While the bird's linear measurements are comparable to Red-tailed Hawks', harriers weigh no more than Cooper's Hawks. This light wing-loading, coupled with the V-shaped wing configuration, gives the harrier a buoyant, tippy quality in flight. Wings up and wide, they rock, unsteadily, even in the lightest of breezes, like a tightrope walker in the sky. In a moderate wind, the birds are positively tipsy!

On windless days, when birds are using thermals or riding faint updrafts on a ridge, they *may* hold their wings flat or even with a slight down-droop, getting the most out of the surface area, and making use of lift that other, heavier birds could not exploit. But the stiffly held, uplifted V-shaped wings are a harrier trademark.

In migration, the bird's course is steady and direct, quite unlike the wandering hunting flight that most observers associate with this species. Except for falcons, harriers rely on powered flight more than other raptors, but harriers do also flap and glide. The wingbeats are regular, measured, lazy, and loping — sometimes continuous, more often given in a series of four to six measured beats followed by a floating glide. The upstroke seems reluctant. The downstroke is deep and precise, but perfunctory, and the last stroke in a series often has a little more *oomph.*

Between wingbeats, harriers just keep gliding along. They don't lose altitude or slow appreciably, and past the point that other birds must stall and fall, they keep gliding — or perhaps floating more than they glide.

Between the bird's lazy, loping wingbeats and the slow, buoyant, dreamlike quality of the glide, the rhythm and pattern of a harrier's flight is unique, a visual fingerprint in the sky. Once familiar with it, an observer can identify Northern Harrier at the limit of detection by its flight alone.

PLUMAGE. Seen from below, the white underparts and ink black wingtips of an adult male harrier distinguish it from all but White-tailed Kite. Juveniles and adult females have a characteristic plumage that is only slightly less distinctive. The wing linings and outer flight feathers of females and juveniles are paler than the secondaries. The result is a two-toned underwing — dark from midwing to the body, pale from midwing to the tip. Juvenile males may sometimes be distinguished from females by their paler flight feathers or silvery hand. Size helps, too.

1. A juvenile Northern Harrier hunting low and slow over a marsh. The cinnamon-infused plumage distinguishes it from adult females. The white rump patch (coupled with the bird's unmistakable lean, lanky profile) is its signature characteristic. Also distinctive is the bold dihedral or V-shaped set of the wings. New Jersey, October KK

2. An adult female Northern Harrier in a glide. The rump patch is invisible from below, but the lanky, long-winged, long-tailed harrier profile remains. In fact, the overall shape is almost idiosyncratic. Heavy, dribbled streaking on the throat and breast distinguish the bird as an adult female. Utah, October JL

3. A soaring adult female Northern Harrier. Note that soaring birds commonly do not fan their tail. This is a particularly lightly marked bird. The more heavily streaked adult female shown in #2 is more typical. Utah, October JL

4. A gliding juvenile Northern Harrier showing no streaking below but a classic profile with wings in a dihedral. Note dark, dusky inner half of wing (arm) and paler hand. Cape May, October JL

5. A soaring adult male Northern Harrier in a tight turn, with tail fully spread. Care to guess why the bird's reverent nickname is "the gray ghost"? Texas, April KK

6. An adult male Northern Harrier in classic hunting flight over open fields. Rufous barring on flanks is variable. Utah, November JL

7. A juvenile Northern Harrier riding an updraft. You can almost imagine the unsteady rocking or tipping motion as this very light, buoyant bird tries to maintain an even keel. Utah, October JL

8. An adult female Northern Harrier flapping. The cadence is lazy, loping, commonly given in an even, unhurried series, followed by a fairly slow glide during which birds teeter or rock unsteadily, even in light winds. The manner and cadence of a harrier's wingbeat is as diagnostic as a fingerprint to those who are familiar with it. Cape May, October CS

9. A juvenile Northern Harrier in a fast glide. The harrier is the great fooler; this bird in this attitude appears remarkably falconlike. Cape May, October CS

10. An adult male Northern Harrier slowing or perhaps stalling over open marsh. The birds cruise and course more than hover, but when they sight or hear prey, short bouts of hovering, close to the marsh, often follow. Alaska, May KK

PUTTING IT TOGETHER: Telling Harriers from the Rest

Northern Harrier

Northern Harrier soars with wings raised in a strong dihedral; when gliding, the bird holds its wings in a modified dihedral.

At first glance, the harrier's distinctive shape and plumage should keep it from being mistaken for any other species of raptor, but for several reasons just the opposite seems true. Harriers are habitually mistaken for other birds of prey. Nobody will ever know how many gliding harriers have been recorded as Peregrines, or how many harriers have been transformed into Cooper's Hawks by observers who could not, after convincing themselves that the bird was *definitely* not a Sharp-shinned Hawk, look beyond the very long rounded tail to notice all the other traits uncharacteristic of Cooper's Hawks.

At all but the most extreme distance, and unless the bird is directly overhead, the exaggerated dihedral of a harrier will be evident. This feature alone should preclude any confusion with falcons and accipiters, which soar on flat wings and glide with wings that may be slightly drooped down or angled down (note, however, that Cooper's Hawks sometimes soar with a very slight dihedral). The tipping or rocking flight also distinguishes harriers from accipiters and falcons and just about all other raptors except for Turkey Vulture and Swainson's Hawk (and, where they occur, White-tailed Kite and Zone-tailed Hawk).

Like harrier, Swainson's Hawk often hunts low, coursing just above the ground. But in shape and in plumage, the birds are more dissimilar than alike.

Mississippi and (particularly) White-tailed kites are probably the only species that might reasonably be confused with a distant harrier. The wingbeats of Mississippi and White-tailed kites are stiffer than a harrier's. When a kite flaps, it accelerates. When it glides, forward momentum slows. Harriers seem to just keep chugging along at the same easy pace whether they are flapping or gliding.

As for distinguishing them from White-tailed Kites, while harriers do hover, they commonly do so close to the ground (less than six feet) and rarely for more than five or ten seconds. Hovering White-tailed Kites are almost always higher (15 to 100 feet) and commonly hover for a minute (particularly when facing into a steady wind). Furthermore, harriers don't kite (that is, they don't hold motionless into the wind without beating their wings). White-tailed Kites do so habitually.

Harriers are most often mistaken for other hawks when they are flying high overhead. Unaware that this ground-hugging hunter migrates as high as any other raptor, observers simply discount the possibility of harrier when studying high-flying birds.

Furthermore, because the harrier has two very distinctive and nearly iconic field marks (white rump patch and dihedral wing configuration), many observers don't bother to familiarize themselves with the bird's other trademark qualities — its lanky profile and idiosyncratic style of flight. When the bird is overhead, and a white

An adult female Northern Harrier from below. It's a view that confounds many observers who are used to seeing the bird hunting close to the ground. KK

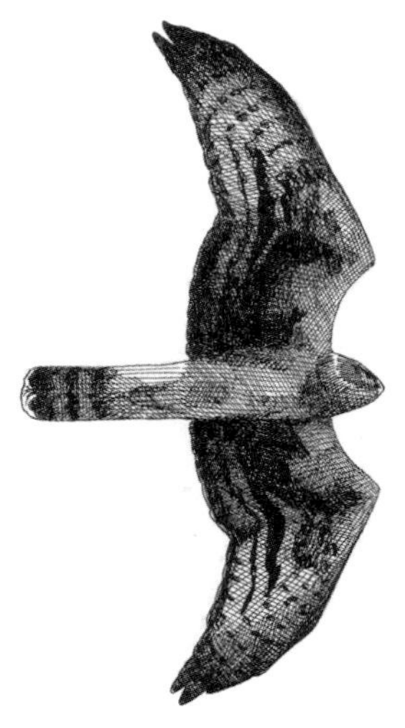

Northern Harrier, high overhead in a steep glide with wingtips swept back, is easily mistaken for a Peregrine Falcon..

rump and dihedral are not visible, observers simply fail to recognize the harrier.

The shape itself is diagnostic; it *is* the field mark! No other raptor in North America shows wings and tail so uniformly, conspicuously, and equally long and narrow. The white underparts of the adult male (with contrasting black wingtips) and the two-toned wing pattern of adult females and juveniles are also distinctive and useful.

Also near idiosyncratic is the harrier's lazy, loping wingbeat. Whether given in a series followed by a short, unsteady glide or in an unbroken, even sequence, the harrier's manner of flight could hardly be more different than the steady, fluid, rippling wingbeats and rapid flight of a Peregrine. In fact, the only thing they might be said to have in common is a direct flight path.

Eagles and Vultures

Big Black Birds

SPECIES
Turkey Vulture, *Cathartes aura*
Black Vulture, *Coragyps atratus*
Bald Eagle, *Haliaeetus leucocephalus*
Golden Eagle, *Aquila chrysaetos*

Only in a book whose focus is the flight identification of raptors could these taxonomically dissimilar species be lumped into a group. In the hawk-watching arena, a playing field whose boundaries push convention, birds drop taxonomic semantics like so many worn flight feathers. It is precisely at the point where the niceties of simple field identification disappear that unifying similarities reign: large size, overall dark coloration, and direct level flight on motionless wings.

Just another way of describing "big black birds."

MIGRATION

Three of the four birds in this group are highly migratory. The fourth, Black Vulture, is sedentary throughout much of its range, but seasonal movements are noted in Mexico and Central America. In the Northeast, where Black Vultures have engaged in a decades-long northern expansion, the number of birds recorded as migrants has increased steadily.

Whether a resident, migrant, or vagrant, Black Vulture remains an element in the big black bird complex over a significant portion of North America — from the Middle Atlantic States, south and west through the old Confederacy to Texas and Mexico, and west to southeastern Arizona.

Turkey Vultures, whose range encompasses almost all of North America north to southern Canada, are a true long-distance migrant. Birds may return to northern portions of their range as early as February, and, in places, movement can be detected into May. Massed movements through Central America and Mexico, and across the

LEFT: Immature, about one-year-old, Bald Eagle. Up close, big black birds present little problem. KK

United States border (involving thousands of birds flying in homogeneous kettles or with other migrating raptors) also begin in February and continue through April.

Fall migration begins as early as August and continues into December with peak numbers appearing in October and November. In Veracruz, Mexico, where up to 2.5 million migrating Turkey Vultures have been recorded in a season, peak flights commonly pass in October.

This thermal-loving species seems acutely hydrophobic, generally shunning water crossings greater than a few miles. Nevertheless, given optimal conditions, Turkey Vultures sometimes cross Delaware Bay, a minimum distance of ten miles, and it is likely that the bird is capable of making even longer crossings.

Bald Eagle migration is complex and protracted, with different age groups and populations behaving differently and adhering to different timetables. Many adults, particularly those breeding south of 40 degrees north latitude, remain in the vicinity of nest sites year-round. Northern birds breeding in the interior of Alaska, Canada, and the northern United States migrate south or to coastal locations before the onset of freezing temperatures, but adults in coastal Maine (for example) remain year-round. What's more, juvenile and nonbreeding subadults (two- to four-year-old birds) are basically nomadic, wandering in response to food needs and availability. Florida juveniles generally head north in February; far northern young generally head south in September.

In a broad sense, fall migration occurs from mid-August through January, with differing peaks in different locations falling between mid-September and early December. Spring migration runs from January through May, with adults migrating earliest and most directly. But young eagles are wandering every month of the year and, in a very real sense, taking a continental perspective. Bald Eagle migration in North America is ongoing, guided, and limited largely by the need to find open, fish-bearing water.

Golden Eagle migration is more proscribed. As a nesting or wintering species, it is sparsely but widely distributed across North America but most numerous in Arctic regions and the mountain West. It is all but absent in the Deep South. Northern birds make great passages to and from their Arctic cliffs. Some birds from northern Canada and Alaska fly all the way to the desert Southwest and northern Mexico. Absent as breeders in the eastern United States since the 1970s, the small number of migrants and wintering birds there hail, presumably, from Labrador and the James Bay region of Quebec.

In autumn, young Golden Eagles depart first, forming the vanguard of the autumn movement; adults follow closely on the heels of the young. In Alaska, peak migration occurs during September; in the intermountain West, it continues from late September into October. In the East, birds may appear as early as September, but most birds pass between mid-October and mid-November.

In spring, adults depart first, some migrating in February, but most in March. Juvenile birds most commonly wait until April and early May. The northernmost breeders arrive from late March to mid-May.

The migration of all the big black birds is thermal oriented (although not dependent). Golden Eagles, in particular—and most particularly in the West—use ridge updrafts extensively during migration. Most migration occurs from midmorning to midafternoon, when thermal production is high.

Eagles and vultures, all species soaring. Turkey Vulture **(A)**. Black Vulture **(B)**. Bald Eagle **(C)**. Golden Eagle **(D)**. Vultures are distinguished by smaller size, bare head, and pale legs. Black Vulture is readily distinguished by stubby wings and short tail. Turkey Vulture is very similar to Golden Eagle, but note two-toned wings and rocking flight. Eagles are best distinguished from one another by plumage. Golden has smaller head, smaller bill, longer tail; wings slightly broader in secondaries, more rounded at tip, with more buteolike shape. Compare position of white patches on wings of these juvenile eagles.

A top view of a Turkey Vulture migrating below a ridge top. Even from above, the signature dihedral is readily seen. NH

Turkey Vulture and Black Vulture commonly migrate in groups or flocks. At some concentration points, most notably Veracruz, Mexico, these may number in the thousands. Bald and Golden eagles usually migrate singly, but pairs of adults are sometimes noted and, at migration concentration points, small groups of half a dozen or more eagles are regularly seen.

IDENTIFICATION

This section might be retitled "How to Turn a Distant Vulture into an Eagle," because in most places that avid hawk watchers gather, eagles are the sought-after bird. Hawk watching is often a matter of sifting though distant vultures (chaff) to find the wheat (eagles).

Plumage, shape, size, and manner of flight are all key considerations in differentiating the big black birds (BBBs), and there are several seasons for this principle, among them *distance*. Big black birds, because of their dark, eye-grabbing plumage and size, are often first noticed at *tremendous* distances, distances well beyond the point where smaller, paler birds become invisible. Often, distant eagles remain distant, straining a hawk watcher's skills to the limits. Identifying distant BBBs means using every hint and clue and being grateful for them.

THE GENERIC BIG BLACK BIRD

This hybridized raptor is a large to massive bird with blunt and broadly proportioned boardlike wings. Its large size is reflected in a slower, more deliberate manner of flight, spare use of wingbeats, and wider turns than smaller raptors. The bird appears dark, almost black. In most cases, they have large gleaming (or dirty) white patches on the wings (particularly the underwings) and the tail. In some cases, the white may extend onto the head or be diffused throughout the body.

VULTURES

Considered to be more closely related to storks than diurnal birds of prey, Black and Turkey vultures share convergent traits with hawks and eagles, including meat-tearing bills and a mastery of flight. Most hawk watchers still accord them honorary raptor status, and their similarity to eagles makes their inclusion in this book axiomatic.

Vultures, both the New World vultures and the Old (which *are* more closely related to birds of prey) are masters of two disciplines: soaring and sanitation. Both the Turkey Vulture and Black Vulture are large and dark onto black, and both feed primarily on carrion. North America boasts one additional vulture species, the California Condor. Rare and geographically localized (and therefore not a top-drawer possibility for most of those trying to tease an identification out of a distant BBB), this species is treated separately.

Black and Turkey vultures have a bare, unfeathered head, an adaptive trait that they share with several Old World vultures. This adaptation is helpful when the birds feed on large carrion — to be more specific, a naked head plunged deep into the interior of a putrefying carcass will not accumulate as much gore as a feathered head.

The Turkey Vulture ranges widely throughout the United States and southern Canada. The population is increasing and expanding northward, keeping pace with global warming, and taking advantage of the proliferation of deer and wasteful predilections of North America's top predator, the automobile.

Most Turkey Vultures vacate the northern reaches of their range in winter (though the wintering range continues to expand northward). Great migrations involving thousands of birds occur in Texas and Central America in spring and fall. Yet, in many places, birds appear to be year-round residents, gathering in communal roosts outside of the breeding season and dispersing in spring and summer. The biological mechanisms governing movement versus nonmovement, or true migration versus dispersal or regional relocation, are not understood.

More southern in distribution, Black Vultures are thought to be less migratory than Turkey Vultures, although movements involving flocks of birds have been reported in Central America, and range increases into the seasonally intemperate Northeast may also, by necessity, prompt some seasonal relocation.

Vultures are highly gregarious. The birds roost together, soar together, and — if a carcass is large enough — feed together. Even during the breeding season, some birds continue to return to established roost sites for the night. Unless persecuted, vultures are not shy around humans. Turkey and Black vultures have become suburbanized, roosting right in suburban neighborhoods. In Central and South America, both species (in particular Black Vulture) are common town and city dwellers.

Excellent soaring birds, vultures commonly wait until thermals begin forming before leaving their perches in the morning and quit the skies as soon as thermal production wanes. In winter, when a shortage of daylight hours and the shallow angle of the sun limit thermal production, vultures often roost on hilltops, letting gravity give them a head start on cold mornings. It might be that the factor limiting the northern expansion for both species is not temperature, but limited foraging time in the night-dominated northern lands.

Turkey Vulture

A Turkey Vulture soaring.

IDENTIFICATION. The Turkey Vulture is large, nearly as big as an eagle. Sexes are similar in size and appearance. From a distance, top and bottom, the bird appears uniformly dark and nearly black (at close range, and in good light, dark brown) except for the flight feathers. Seen from below, the entire outer and trailing edge of the wing shines silver gray, contrasting markedly with the all-black underparts and somewhat grayer tail. Adults and immatures differ in head and bill color. Adults have a naked red head and bright ivory bill; juveniles have a dull gray head and gray bill (as do Black Vultures). *Caution: Just because a vulture doesn't have an all-red head doesn't mean it is a Black Vulture.*

Subadult (second-year) Turkey Vultures have a reddish head but a bicolored bill — ivory near the base, dark at the tip.

IN FLIGHT. The wings are long, broad, and planklike, but not devoid of bulges or contours along the trailing edge. The feathers at the wingtips are deeply slotted, resembling widespread fingers. The tail is fairly long but wide — very rectangular when closed, *which it is most of the time.* Even when soaring, a Turkey Vulture commonly keeps its tail wholly or mostly closed, not fanned.

The head is diminutive, ridiculously small. Even with its dimensions exaggerated by a fluffy, feathered neck ruff, at a distance, the head on a Turkey Vulture disappears. The bird seems nothing but wings and tail.

Turkey Vultures hold their wings sharply uplifted in a bold dihedral. In a soar, the wingtips turn up, forming a U-shape; in a glide, the lines straighten, giving birds an exaggerated V. Several other species of raptors hold their wings this way, including Golden Eagle, Northern Harrier, Swainson's Hawk, Ferruginous Hawk, White-tailed Hawk,

Turkey Vulture, upperside. All feathers brownish black; some pale brownish coverts, contrasting with darker flight feathers and coverts, are also visible at rest and can be helpful in separating perched vultures; Black Vulture is plain coal black.

Turkey Vulture, underside. Long broad wings held in pronounced dihedral; broad, fingered wingtips; fairly long tail; tiny head. Black body and underwing coverts contrast with silvery flight feathers. Upper birds show underwings in shadow, lower birds with wings catching direct sunlight.

One Black Vulture (second from right) with three Turkey Vultures, showing differences in tail length and wing proportions.

and Zone-tailed Hawk. These birds, however, do not usually hold their wings as acutely upturned as Turkey Vulture does.

With the exception of Zone-tailed Hawk and, to a lesser degree, Northern Harrier, only Turkey Vulture habitually rocks or teeters side to side in flight. Some buteo species teeter somewhat, among them Swainson's Hawk and Ferruginous Hawk. But the rocking motion on these species seems measured, controlled. Turkey Vultures *teeter* unsteadily even in windless conditions, *wobble* when it's breezy, and *reel* drunkenly in strong winds.

This chronically tippy flight exposes the vulture's silver underwings. At every tip up, the underwings *flash* silver, like a mirror catching the sun. The effect is dramatic, visible at great distances and virtually diagnostic, a blinking message that reads: "Hi, Turkey Vulture here!"

Once airborne, Turkey Vultures flap little. In calm air, their flight is buoyant and acrobatic, but somewhat amateurish — stiff, hesitant, and (of course) unsteady. When Turkey Vultures do flap, the wingbeat is heavy, labored, exaggerated, and deep. A Turkey Vulture also has the curious habit of waving its hands (i.e., drooping its wingtips, then straightening them out) in a sort of mock flap. The bird's arms remain rigid and uplifted. The hands simply wilt, then right themselves.

Golden Eagle is probably the bird most likely to be confused with Turkey Vulture. Golden Eagles are large, dark, long tailed, and proportionately short-headed. Golden Eagles also soar and glide with their wings held in a dihedral (although not as exaggerated a dihedral as in Turkey Vulture). But Golden Eagle does not rock in flight, and its head, while smaller than the head of Bald Eagle, is nevertheless prominent. While the head of a distant or high-soaring Turkey Vulture might disappear, on Golden Eagle it never does.

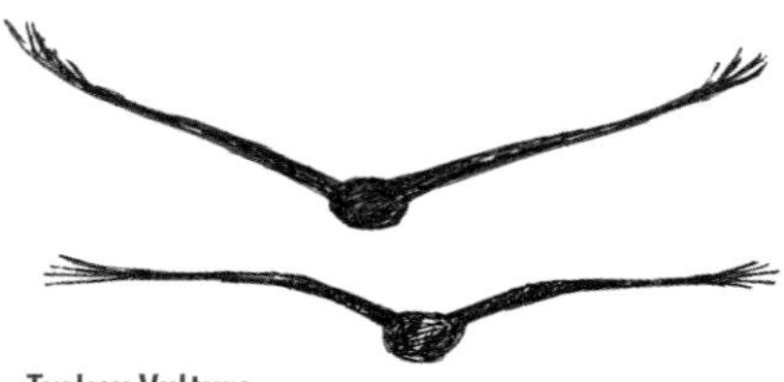

Turkey Vulture

Turkey Vulture soars with wings in a strong and upswept dihedral; it glides with wings in a modified dihedral with the angle closer to the body than in other species.

1. An adult Turkey Vulture — a big black bird with a face only a hawk watcher could love. This image is classic, showing a bird in a full soar. Note the red head and yellow bill of an adult, the pronounced dihedral, the two-toned underwings, and the long (and in this case) fanned tail. Unlike this banking bird, many Turkey Vultures soar with tail closed. The tail still appears broad, just not fanned (also see #2). New Jersey, March CS

2. An adult Turkey Vulture in a floating glide. Note the pronounced dihedral, silvery flight feathers, and deeply slotted outer primaries. Note the color. Given exceptional lighting conditions, the birds appear dark brown, not black. New Jersey, December CS

3. A juvenile (gray head) Turkey Vulture, soaring. Because of their relatively small heads, high-flying Turkey Vultures appear headless, all wing and tail. Cape May, October CS

4. The upperside of a soaring juvenile Turkey Vulture (same bird as #3). Marked as juvenile by the clean pale scalloping on the wing coverts. Note that Turkey Vulture's silvery underwings do not show from above. Black Vultures, however, show a whitish patch at the wingtips both above and below. CS

5. An adult Turkey Vulture, here in a fast glide. While the image is frozen, you can bet that when this photo was taken the bird was teetering or wobbling. Unsteady flight, whether the bird is soaring or gliding, and whether it is windy or calm, is a signature trait of Turkey Vulture. Cape May, October CS

6. Turkey Vultures streaming during migration. Their glide is slow and floating. They are all on the same glide path and compass bearing, but their constant unsteady wobbling has two of the birds out of kilter with the rest. Note how the contrast between the leading and trailing edge of the wing diminishes on wings not catching direct sunlight. Note, too, the no-headed appearance; eagles don't appear headless, and even Zone-tailed Hawk shows a broader, more prominent head. Cape May, October CS

Black Vulture

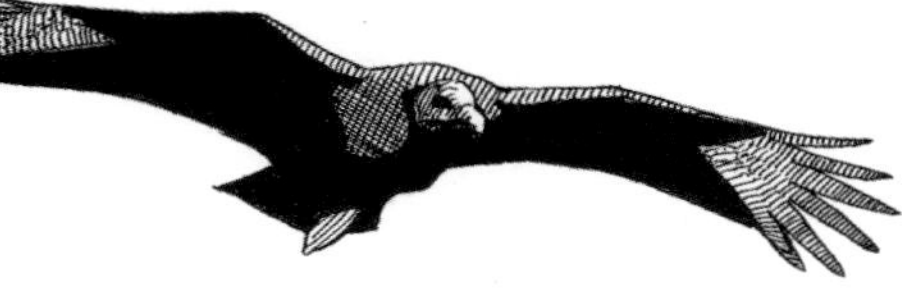
A Black Vulture, head-on.

IDENTIFICATION. The Black Vulture bears some resemblance to the Turkey Vulture, but at a distance it is far more likely to be mistaken for an eagle. Like other species in this grouping, Black Vultures are overall dark, almost wholly black, except for the naked gray head and prominent grayish white to buffy white (not silver) patches on both wingtips. This feature is prominent on the underwing but less prominent on the upperwing surface. Except in Florida and Texas, where Northern Caracara also occurs, a large dark raptor with obvious white wingtips is automatically a Black Vulture. Plumage does not vary with age or sex.

IN FLIGHT. Black Vulture is the smallest member of this foursome. It is a little more than two-thirds the size of Turkey Vulture, about half the size of either eagle species — just about Red-tailed Hawk–sized. The wings are short and stocky. The tail is ridiculously short, stubby, and wedge-shaped; it's always fanned.

So short is the tail that the bird's feet (which appear whitish thanks to a layer of caked excrement) can sometimes be seen trailing behind the tail tip (but this

Black Vulture, underside. Short and squat. Wings and tail relatively much shorter than on Turkey Vulture; shape more angular, head rounder, bill thinner. Uniformly coal black except for six silvery outer primaries.

Black Vulture, upperside. Uniformly black except for white shafts of outer six primaries; compare browner Turkey Vulture.

characteristic is not dependable as a field mark). At a distance, birds appear tailless and all wing — the tail is simply lost in the wing.

The naked head and neck protrude no more than the head of a Turkey Vulture but, because of the Black Vulture's overall stockiness, the head appears larger, or at least more prominent. The head and tail of Black Vulture are equal in length. The tail of Turkey Vulture is about three times the length of its head.

Putting it all together, in flight, a Black Vulture looks like a stocky, heavy-necked, short-tailed eagle, a compact coffee table of a bird. Given a side view, the bird appears front-heavy; the body seems to hang or dangle beneath the wing.

When soaring, a Black Vulture's wings are held stiff and flat, or with a slight dihedral. In a glide, wings are mostly flat or very slightly down-curved. Confusion with Turkey Vulture is most unlikely — in fact, and as stated before, Black Vulture is far more easily confused with an eagle.

In flight, Black Vultures lack the tippy buoyancy of a Turkey Vulture or the steady stateliness of an eagle. The flight is generally steady, but movements seem stiff and amateurish. The bird holds its wings tensely perpendicular to the body and holds itself aloft in a fashion that brings to mind the image of someone riding a bicycle for the first time (and hanging on for dear life). Black Vulture flaps a great deal more than does Turkey Vulture. The wingbeats are always given in a series — quick, choppy wingbeats that have a hurried quality about them. In a mixed flock of vultures, Black Vultures look like a tagalong among older siblings, doing their best to keep up. Also, Black Vultures sometimes soar, for short periods, with their legs down; Turkey Vultures almost never do.

Black Vultures do not rock in flight and have little or no dihedral. They are also darker, blacker than Turkey Vultures — a difference that can be noted when the birds are seen soaring together, but one that is hardly needed since size, shape, and manner of flight distinguish them clearly.

Black Vultures are prone to flying abreast, as opposed to strung out like Turkey Vultures. Black Vultures also frequently fly at greater altitudes than do foraging Turkey Vultures.

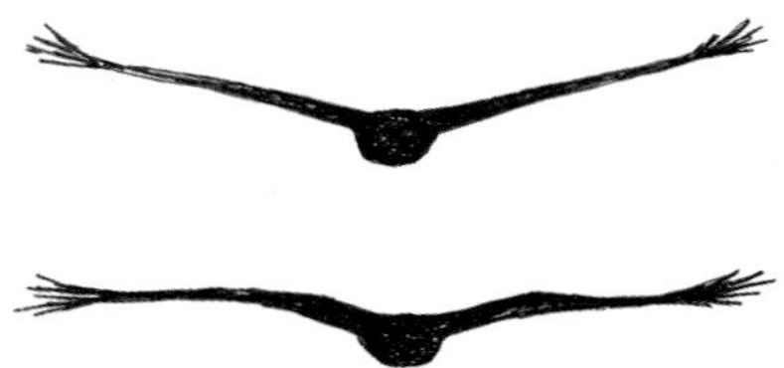

Black Vulture

Black Vulture soars with wings raised slightly and glides with wings nearly flat.

1. An adult Black Vulture in a full soar (juveniles are essentially similar). They really are black! Turkey Vultures are dark brown. Note: Black Vultures usually fly with tail fanned. Turkey Vultures habitually fly (and soar) with their tail wholly or mostly closed. Florida, January KK

2. A soaring Black Vulture showing short stocky wings thrust slightly forward, short triangular tail, and triangular head projecting about as far ahead of the wing as the tail projects behind. White wingtips are sometimes less prominent than shown here. Overall, a very blocky, compact bird. Cape May, November CS

3. A Black Vulture in a glide but still showing the equidistant head and tail characteristic of this species. In a very acute glide, seen head-on, the wings jut up at the wrist and angle down along the arm. You get a sense of this here. Cape May, October CS

4. A Turkey Vulture ahead and a Black Vulture behind, in a shallow glide. This is typical when the species fly together: Turkey Vultures commonly lead. The different overall shapes and relative proportions are very apparent in this shot. Cape May, October CS

5. A Black Vulture below and a Turkey Vulture above. In a full soar, Black Vultures appear tailless (the tail fuses with the trailing edge of the wing). Turkey Vultures appear long tailed (and short headed) no matter what. Cape May, November CS

EAGLES

Except for California Condor, Bald and Golden eagles are the largest birds of prey in North America. Regard for the birds has been two-sided. On the one hand, eagles are admired for their physical prowess and hunting skills. The Bald Eagle serves as the emblematic symbol of the United States, and the Golden Eagle has found its way onto more than one nation's (and empire's) standards. But the birds also have their detractors. The Bald Eagle, whose diet includes the remains of winter-killed or hydroelectric-damaged fish, has been called a scavenger and a carrion eater. The Golden Eagle has been persecuted by sheepherders because of real (but exaggerated) claims of eagle depredation upon stock.

Two other eagle species have occurred in North America, both placed in the same genus as Bald Eagle: White-tailed Eagle (*Haliaeetus albicilla*) and Steller's Sea Eagle (*H. pelagicus*). The former, a large fishing eagle that has a white-laced head and white tail, is resident in Greenland and Asia. There are old records for this species on the Atlantic side of North America, but appearances by Asian birds in coastal Alaska are now more likely. Into the late 1980s, a pair of White-tailed Eagles nested on Attu, the westernmost island in the Aleutian Island chain.

The very large, handsome Steller's Sea Eagle is a resident of northeastern Asia; it winters south to Japan and occasionally wanders to the Aleutians, Kodiak Island, and, rarely, the North American mainland. A single adult bird took up summer residence not far from Juneau in the 1990s.

But for purposes of flight identification, White-tailed Eagle and Steller's are long shots, possibilities not to be counted on or variables not to be considered when identifying a distant BBB over most of North America.

Few birds can be identified as far or high as an adult Bald Eagle. Even fewer birds are as iconic. New Jersey, February CS

Bald Eagle

In North America, the Bald Eagle is the more widely distributed of our two native eagles. Unlike the Golden, which occurs in Europe and Asia, the Bald Eagle is wholly confined to North America north of Mexico. Its range, broadly speaking, covers all of North America south of the tree line (although, in truth, the bird is uncommon away from coastal areas, large inland lakes, and rivers). The Bald Eagle is one of the fish-eating eagles, a description that is neither wholly accurate nor fair because the bird is an accomplished hunter as well as a fisherman. Fish, captured as prey, appropriated as carrion, or pirated from Osprey (or another Bald Eagle), does constitute a large portion of the Bald Eagle's diet, but waterfowl are also a large part of the bird's fare. Using an adroit combination of herding and powered flight, Bald Eagles simply overtake ducks and geese in flight. Winter-weakened and crippled birds probably bear the brunt of this harvest, but accomplished adults have no apparent difficulty catching healthy waterfowl.

Rather than *scavenger,* the word *forager* more aptly describes the Bald Eagle's food-gathering focus and technique. Perch hunting is a practiced art with Bald Eagles. One will sit for hours on a strategic perch, watching a river for whatever fortune the waters ferry its way, or simply waiting for hunger to propel it to action. Humans might assume this practiced inactivity is a display of laziness, overlooking the fact that sitting is energy conservation in effect (not to mention a testimony to the bird's foraging skills). Birds whose energy needs are met can afford to sit. Hungry birds must search.

Bald Eagle, underside. First year **(A)**. Adult **(B)**. Wing shape slightly less buteolike than on Golden Eagle but very similar; head larger, tail slightly shorter. Adults have thinner, straighter wings than immatures. Adult unmistakable; immatures extremely variable (see p. 207). Individual illustrated is a more or less typical first-year bird, with dark body and patches of white on underwing coverts, mainly on axillaries and along median coverts.

Bald Eagles of different ages. One-year-old, dark extreme **(A)**. Two-year-old, typical **(B)**. Two-year-old, white extreme **(C)**. Four-year-old **(D)**. Dark-bellied first-year bird shown here **(A)** is as dark as Bald Eagles get, in the plumage most difficult to distinguish from Golden Eagle's. By age two, birds are white bellied, as in **(B)**, and some become mostly white, as in **(C)**, suggesting Osprey, before molting to dark-bodied adult-type plumage by fourth year. Bird **(D)** is adultlike, with narrow, straight-edged wings and all-dark body, but retains dark line through eye and dark tip on tail.

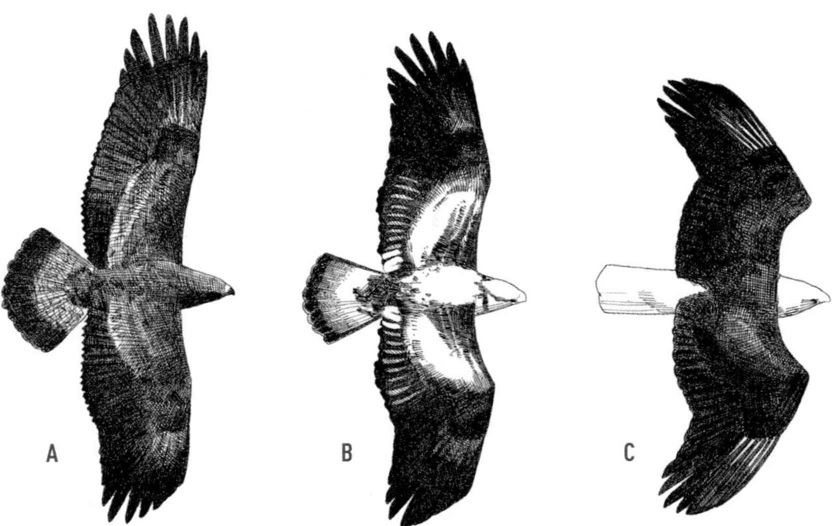

Bald Eagle, upperside. First year **(A)**. White extreme, second year **(B)**. Adult **(C)**. Progression of plumages resembles that on underside. First-year birds all-dark above, with pale base of tail; light patches on wing coverts appear with wear. All white-bellied birds **(B)** have a triangular white patch on the upper back; this individual is the white extreme. Adults are very dark brown except for pale primary shafts and white head, tail, and rump.

IDENTIFICATION. The Bald Eagle is a massive, plank-winged, darkly plumaged raptor that is larger than any buteo or vulture and is comparable only to Golden Eagle in size. Individuals vary considerably in size (females may be much larger than males), but even the smallest Bald Eagle will seem immense compared to other raptors (condor excepted).

Birds four years old and older show the white head that is its namesake characteristic and a white tail. Immature Bald Eagles (1–3 year-old birds) lack these easily noted characteristics and, as a result, may be confused with Golden Eagles, vultures, and some buteos.

Immature Bald Eagles are highly variable, from mostly dark individuals that show limited amounts of white in the plumage to those that show lavish amounts of white and appear overall mottled or mocha-colored. During the first four years of its life, a Bald Eagle undergoes ongoing changes in plumage — a general, overall *whitening* process that culminates with the white coalescing in the head and tail, leaving the rest of the bird uniformly dark.

First-year (juvenile) Bald Eagles are blackish brown, appearing all black at a distance or showing two-toned upperparts (light brown back and upperwing coverts, dark brown flight feathers). The underwing linings, the axillars (or wingpits), and the base of the tail are white. Except for some spare white mottling on the flight feathers or the belly, the balance of the bird is dark.

From a distance, juvenile birds simply look uniformly dark with pale wing linings. Tails are variable — some are all dark with white limited to the base; some are mostly white (*dirty* white) with a ragged dark band at the tip. Insofar as first-year birds, whose ranks have yet to be whittled down by mortality, make up a large portion of the eagle population, *particularly in fall,* this plumage is the one most commonly seen by autumn hawk watchers.

Second- and third-year Bald Eagles (advanced immatures) show increasing amounts of white all over the body and in the wings. White underwing linings and axillars remain prominent, but advanced immatures develop lavish amounts of white on the back, the flight feathers, and particularly the belly. Head and breast are generally dark, although the head of a third-year bird begins to whiten, commonly showing a dark Osprey-like eye stripe. The tail is variable but usually ringed — a dirty white tail with a well-defined dark terminal band.

It is this advanced-immature plumage that is most characteristic of immature Bald Eagles seen in spring, and it is these that will dominate until the new class of juvenile Bald Eagles is recruited into the ranks (with some Florida birds fledging as early as January and some Alaska birds not leaving the nest until August). This protracted nesting period, coupled with plumage developmental variation between individuals, makes the assigning of age or year classes to advanced immatures (often denoted Basic I and Basic II) problematic.

Suffice it to say that advanced immatures (post-juvenile, pre-definitive) look ragged, disheveled, unkempt. At a distance, owing to the lavish amounts of white feathering infused into the plumage, birds may appear overall pale brown, mocha, or even latte-colored.

Four-year-old eagles begin to assume classic adult characteristics, although the white tail may still show traces of a black tip, and the face may be marked by a dark eye line. Adult plumage is usually wholly assumed between the fifth and sixth years.

The preceding discussion is perhaps more detailed than necessary. For species identification purposes, all an observer needs to know is this:

- If an eagle-sized raptor is all dark except for a gleaming white head and a gleaming white tail, it is a *Bald Eagle.*
- If an eagle-sized raptor is all dark except for white or pale underwing linings, it is a *Bald Eagle.*
- If a large raptor has lavish amounts of dirty white and buffy mottling all over the wings and body, particularly the belly, it is a *Bald Eagle.*

An adult Bald Eagle, soaring and turning away.

Except for their golden-glazed crowns and a hint of white at the base of the tail, adult Golden Eagles are uniformly dark. Juvenile Golden Eagles are, likewise, overall dark except for well-defined and gleaming white patches *on the base of the flight feathers and a gleaming white tail (particularly above) tipped with a broad, crisply defined dark band.*

IN FLIGHT. The tail of most first-year Bald Eagles is dark for two-thirds to three-quarters of its length, and white only at the base. Advanced immature Bald Eagles have a ringed tail that is very similar to juvenile or second-year Golden Eagles, but the tail of young Bald Eagles is not so white, and the terminal band not so crisp.

More important, the head of a Bald Eagle is large and projecting. The bill is prominent — as much as one-third of the total projection. On adult Bald Eagles, the head and tail are almost of equal length. Set on long planklike wings, the bird looks symmetrical and balanced.

Immature Bald Eagles (particularly juveniles) have longer tails, half again as long as the head. Nevertheless, immature Bald Eagles still appear very big headed. Without much exaggeration, the head and tail on immatures appear to be close to equal length.

The tail of a Golden Eagle, however, is noticeably longer than the bird's head projection, about twice the length of the head. This long-tailed, small-headed proportion gives Golden Eagles a very buteolike appearance. A Bald Eagle doesn't look like a buteo.

In a soar, the wings of an adult Bald Eagle look incredibly long and uniformly wide, planklike, a little too narrow for the rest of the bird. The lines are very straight, almost devoid of bumps or bulges. Immature eagles have broader and more tapered wings that still appear planklike — maybe doorlike!

When soaring, the bird holds its wings at a right angle to the body and straight out. Most often, the wings are flat throughout their length, sometimes showing a very slight dihedral (particularly in adults). In very calm conditions, and when the bird is gliding, the wings may droop modestly and along their length, suggesting an Osprey.

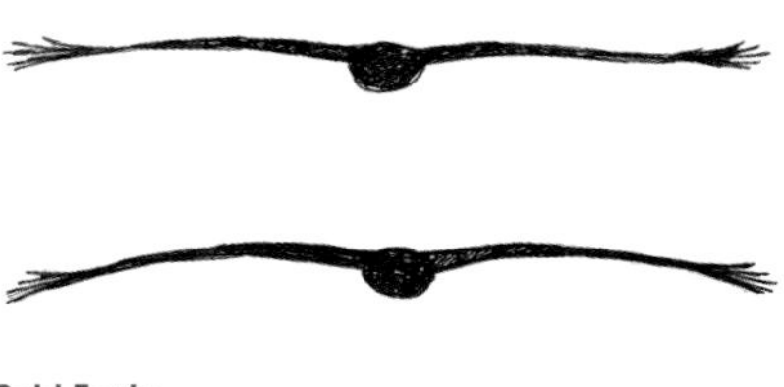

Bald Eagle

Bald Eagle soars with wings flat or slightly drooped; it glides with wings flat but tips distinctly drooped.

In a soar or in a glide, the movement is steady, stately, deliberate, and even-keeled. Turns are wide and slow. Soaring eagles hardly ever flap.

In level flight, the wingbeat is slow, heavy, measured, somewhat stiff, and surprisingly shallow — describing or tracing an arch of about 40 degrees. Except for birds just leaving a perch or accelerating to capture prey, wingbeats are not deep. Instead, wingbeats project a sense of balance, symmetry, and precision, with the wings rising above the body as much as they fall below, and as much effort apportioned to the upstroke as to the down. By contrast, on most raptors, wings appear to barely rise above the body, and most of the effort is directed below.

Bald Eagles are adept at powered flight and may flap frequently, even continuously, as they span the sky from horizon to horizon. But in migration, if thermals are strong or updrafts good, the birds may never flap at all.

1. An advanced immature Bald Eagle showing a white belly and Osprey-like eye line. The flight of Bald Eagle is fast and direct; wingbeats are powerful with as much (or more) upstroke rising above the body as downstroke falling below. Utah, November JL

2. An adult Bald Eagle in a glide. Unmistakable, yes, but take note of the size of the head relative to the length of the tail. On adult Bald Eagles, there is about as much head projecting in front of the wing as there is tail behind. Juveniles appear longer tailed but show similar head projection. New Jersey, January CS

3. A soaring adult Bald Eagle showing the planklike wings (and near equal projection of head and tail). New Jersey, January CS

4. A worn juvenile Bald Eagle transitioning into its white-belly plumage. Note the big head (and massive bill), dull white wing linings, and dirty white infused in the tail. Alaska, May KK

5. A fairly typical juvenile Bald Eagle in a glide or perhaps riding an updraft off a ridge. Note the dirty white underwing linings contrasting with the whiter wingpits and dirty white touches in the tail. New Jersey, November CS

6. A juvenile Bald Eagle with more tuck and in a faster glide than #5. This individual's tail contains more white and shows a semblance of a tail band. But the white in the tail of juvenile Golden Eagles is gleaming white, and the underwings would never show such extensive white. New Jersey, February CS

7. A juvenile Bald Eagle from above. While overall dark, upperparts are commonly two-toned. As they age, young eagles become progressively whiter on the back (with the white often taking the shape of a triangle) and, at a distance, the mix of dark and dirty white feathers gives birds a mocha-colored appearance. Utah, November JL

8. A soaring advanced immature Bald Eagle showing lavish amounts of white (most notably on the belly) and the Osprey-like eye pattern seen on advanced birds. Many subadult birds appear even more ragged or disheveled than this individual. Utah, December JL

9. Young Bald Eagles are highly variable. Compare this juvenile bird (above) with the third-year bird (below). Note the more pointed tips on the secondaries of the juvenile, as well as its longer tail. Alaska, May KK

10. An advanced immature Bald Eagle banking. The white on the body diminishes as birds approach adult plumage. Utah, November JL

11. An adult Bald Eagle in a full soar showing characteristic planklike wings. Cape May, November CS

Golden Eagle

If the Bald Eagle is a forager, the Golden Eagle is a hunter. More restricted than Bald Eagle in North America, Golden Eagle is common and widespread but sparsely distributed across much of the West — from northern and western Alaska, east across much of the Canadian Barren Lands, and south across the western United States into northern Mexico. In winter, birds vacate the northern half of their breeding range, withdrawing to southern Canada, southern portions of the birds' range, and east into the prairies.

Golden Eagles are much less common across the central and eastern United States and southeastern Canada. They breed in northern Ontario, northern Quebec, northern Newfoundland, and northern New Brunswick. As recently as 1999, two pairs nested in Maine, and two breeding efforts were spawned by reintroduction programs in Tennessee and Georgia. In winter, a relatively small number of Golden Eagles winter widely across the eastern half of the country (most notably the southern Appalachians), excluding the Great Lakes, northern New England, and the Florida peninsula.

Golden Eagles are not dedicated flatlanders, although birds do winter in coastal marshes of the East and the prairie and sage land of the West. Cliff nesters, mountain hunters, and ridge migrants, Golden Eagles prefer terrain that is at odds with the horizon. Though not averse to water, they are not particularly drawn to it (unlike the Bald Eagle, which is a bird of coasts, lakes, and rivers). The Golden Eagle is primarily a bird of dry and sometimes arid mountainous and hilly terrain.

Prey is predominantly medium-sized mammals. In the Arctic, the most favored prey is Arctic ground squirrel. Over much of the West, birds target jackrabbits and other rabbits and hares. In the East, a higher percentage of waterfowl and water birds figure in the Golden Eagle's diet (including geese and Great Blue Herons). Adult Golden Eagles regularly take prey as large as foxes, coyotes, and young mountain sheep.

Eagles hunt from aloft, although they might first spot prey from a strategic perch (one that offers elevation, good visibility, and proximity to thermals or updrafts so birds can gain elevation quickly). The stoop of a Golden Eagle, immortal-

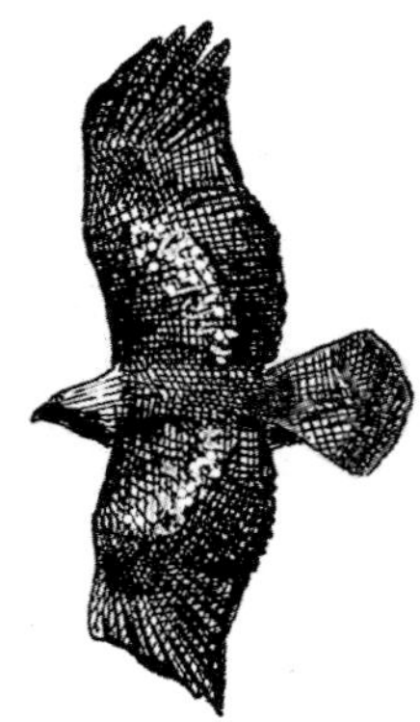

Adult (left) and juvenile Golden Eagle. Adult has dark tail and mottled pale upperwing coverts. Most juveniles do not show white on upperwing, but do show white base of tail and uniformly dark wing coverts.

Golden Eagle, underside. First year **(A)**. First year with less white **(B)**. Adult **(C)**. Compared with Bald Eagle, note smaller head, longer tail, slightly more buteolike shape. As in Bald Eagle, juveniles appear broader winged and longer tailed than adults. Typical first-year birds have large white patches at base of primaries and base of tail; some may lack white in wings **(B)**. Individuals may occasionally show white spots on body or underwing coverts as shown in **(B)**; compare the more extensive and regular pattern of white on immature Bald Eagles.

Golden Eagle, upperside. Juvenile **(A)**. Adult **(B)**. Golden hackles are conspicuous in all plumages. Immature shows small or no white patches on wings, bright white base of tail, rich brown back and wing coverts. Adult has dark tail with pale gray base, primaries and secondaries gray with dark barring. Old and worn coverts produce light patches and bars on wings.

ized in Alfred, Lord Tennyson's poem, is one of nature's most iconic displays. In a stoop or in a powered, level approach, birds prefer to hunt into the wind as much for the maneuvering advantage it affords the attacker as for the disadvantage it levies upon fleeing prey.

In hilly country, Golden Eagles also hunt by coursing low along hilltops, using updrafts and concealing topography to best advantage.

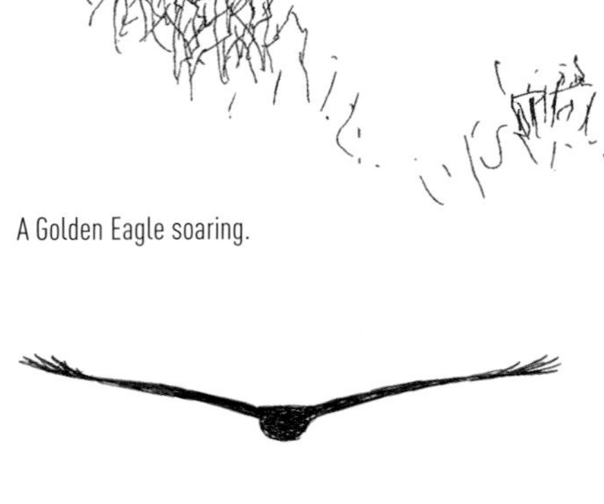

A Golden Eagle soaring.

Golden Eagle

Golden Eagle soars with wings in a slight dihedral and glides with wings essentially flat.

IDENTIFICATION. Golden Eagle is very large and mostly overall dark. In appearance and shape, it is most similar to juvenile Bald Eagle. Adult Golden Eagle is mostly blackish brown, often with traces of gray or some white at the base of the tail. In flight (and when perched), the paler upperwing coverts form a tawny or golden bar along the upperwing. From below, wings are slightly two-toned. Darker underwing linings and a dark trailing edge to the wing frame paler grayish flight feathers — a pale rendering of the bold, classic immature Golden Eagle wing pattern described below.

Juvenile and immature Goldens are likewise blackish brown but often boast gleaming white patches on the wings and a bright white tail broadly trimmed with a well-defined dark terminal tail band. Wing patches, centralized at the base of the inner primaries, may be large and prominent (encompassing one-third or more of the wing). Showing mostly on the underwing, the white patch on the wings of very boldly marked individuals may also appear on the upperwing. The wing patches may also be small (occasionally absent), forming a small rectangular patch restricted to the base of three or four flight feathers and visible *only* from below.

Caution: On some advanced immature Golden Eagles, and perhaps some adults, limited amounts of white on the wings may appear as a narrow or broken white line along the base of the flight feathers and running the length of the wing. It can be mistaken for the pale wing linings of immature Bald Eagle.

Both adults and juveniles have iridescent gold feathers on the head and nape. The amount of gold varies, but on most individuals it is extensive and conspicuous — so much so that at a distance, very pale-headed birds might be mistaken for adult Bald Eagles. Most eastern Goldens are conspicuously gilded. Some western birds show only touches of gold on the head.

IN FLIGHT. In size and shape, the birds are unmistakably eagles but the overall shape is more buteoine than the big-headed, plank-winged Bald Eagle. A Golden's wings are not just long but broad. The trailing edge shows an S-shaped contour, the

Turkey Vulture (left) and juvenile Golden Eagle soaring, similar except for the eagle's white wing and tail markings, larger feathered head, and more buteolike proportions.

product of a hand that appears smaller than Bald Eagle's, plus bulging secondaries, and a wing that pinches where it joins the body.

Adults have shorter, blunter, wide-handed, more paddle-shaped wings so the trailing edge forms a more compact S. The wings of juvenile Goldens appear longer and narrower (particularly in the hand).

Golden Eagle's head is comparatively small, as is the bill. The tail, by comparison, is quite long, twice the length of the head. Bald Eagles have a large projecting head and tail of near equal length.

Viewed as a whole, a Golden Eagle, with its long, broad, shapely wings, its small head, and its long tail, greatly resembles an overgrown dark buteo — something Bald Eagle hardly ever does.

When the bird is soaring or gliding, it holds its wings flat or in a moderate dihedral and the rise along the wing is gentle and sweeping (not abrupt as in Turkey Vulture). Still, seen head-on, Golden Eagles are frequently mistaken for Turkey Vultures on ridge sites. Birds in a tucked glide or coasting along a strong updraft sometimes hold their wings in a slightly raised bow — but birds showing this configuration also appear distinctly short winged.

The wingbeat of a Golden Eagle is similar to that of a Bald Eagle: slow, even, measured, deliberate to the point of being ponderous, and fairly shallow. As in Bald Eagle, the rise of the upstroke and fall of the downstroke are equidistant. However, Golden Eagle, the committed ridge rider, prefers gliding to powered flight. Also, the wingbeat of Golden seems slightly more supple, not as stiff and planklike as Bald Eagle's — again, more suggestive of a very large dark buteo.

As an aerial hunter, Golden Eagle spends considerable time aloft, often very high. It can kite very briefly (more a controlled stall) but does not hover (as is common with Rough-legged Hawk).

Golden Eagle (left) and two Turkey Vultures gliding and going away. The more substantial tail and broader, more buteolike wings of the eagle are apparent.

1. An adult Golden Eagle in a shallow glide along a ridge top. Characteristics to note: all-dark plumage with the hint of a golden nape, and small head and bill. The tail appears two to three times the length of the head. Utah, October JL

2. The dorsal view of an adult Golden Eagle. Golden nape contrasts with the mostly dark upperparts. Pale mottling on the upperwing coverts fuses into a golden-colored carpal bar at a distance. Utah, September JL

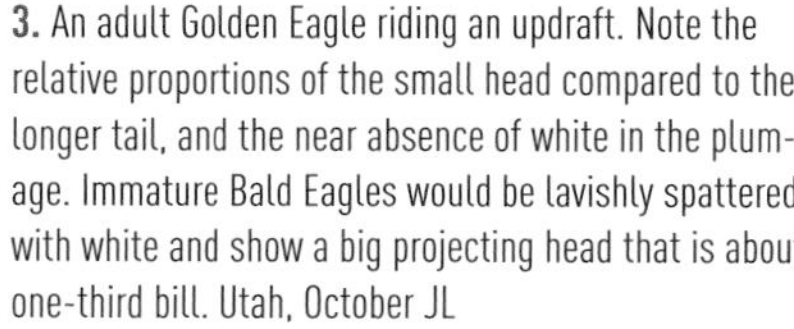

3. An adult Golden Eagle riding an updraft. Note the relative proportions of the small head compared to the longer tail, and the near absence of white in the plumage. Immature Bald Eagles would be lavishly spattered with white and show a big projecting head that is about one-third bill. Utah, October JL

4. An adult Golden Eagle gliding high overhead. With its smaller head, longer tail, and showing this wing configuration (with fairly straight-cut trailing edge), the Golden Eagle more closely resembles large buteos than does Bald Eagle. Compare this image to Bald Eagle #6. Utah, October JL

5. A juvenile Golden Eagle showing classic shape and classic field marks — white patches at the base of the primaries; wide, even dark band trimming the bright white tail. Utah, November JL

6. A juvenile Golden Eagle in a full soar. Note not only the placement of the white, but the brilliance. The white on juvenile Bald Eagle is dingy or dirty white. Also, the body of Golden Eagle tends to be overall darker, blacker than many young Bald Eagles, particularly subadults, which range from overall mocha-colored to cream-colored. Note also wings tapered in at the body, creating a slightly paddle-shaped profile, a shape noticed more on juveniles than adults. Compare this to Bald Eagle's more planklike wings. Nevada, September JL

7. A soaring juvenile Golden Eagle, from above. The dihedral is manifest here, as are the classic field marks. Note also the gleaming white of the tail. The small white patches bleeding through the wings are often lacking. Utah, September JL

8. Riding a mountain updraft, an incoming adult Bal— No, make that Golden Eagle. At times the golden hackles can be so brilliant that an oncoming bird may first be mistaken for an adult Bald Eagle. But note the dihedral. Bald Eagles are flat winged in a soar and glide with wings slightly drooped. Utah, October JL

9. An adult Golden Eagle, soaring high overhead. Note the small head, long tail, and sculpted wings. Compare this shape to Bald Eagle #3. A few adult Goldens show the leading edge of the wing tinged in gold. Arizona, January CS

PUTTING IT TOGETHER: Telling Big Black Birds Apart

When an eagle appears at the limit of conjecture, it gives the impression of being, simply, *a big black bird.* Because an eagle is larger than lesser raptors, and because its blackness contrasts with the clouds or sky, it will be detected at distances well beyond the point that other birds melt into the ether. Vast distance masks any sense of forward movement. The bird (and at this point you may still be trying to certify that it is a bird) will seem to move with infinite slowness or not at all. Any barely moving black dot on the horizon should trigger heightened interest.

Occasionally, in poor light or on an overcast day, a large Red-tailed Hawk or Ferruginous Hawk may appear eagle-sized. As a rule, it is far easier to turn Red-taileds into eagles than to turn eagles (or vultures) into buteos.

When the bird is flapping, the slow, ponderous, deliberate cadence of an eagle's wingbeat is discernible at almost unimaginable distances. The flap of a Black Vulture is quick and choppy, not heavy and deliberate. Turkey Vultures seem to flap only as a last resort and rarely flap steadily, without an intervening glide, as eagles (particularly Bald Eagles) do. Also, the flap of an eagle is shallow, while that of a Vulture is theatrically exaggerated and down pushing.

On set wings, either soaring or gliding, Turkey Vultures hold their wings in an acute dihedral and rock unsteadily. Eagles do not rock in flight and, in fact, can maintain an even keel under the highest winds that any observer is likely to encounter. Since motion projects itself at distances beyond the point where plumage characteristics and relative proportions are discernible, this is often the first tangible clue to an approaching bird's identity. Accordingly, if you are looking at a big black bird and it rocks in flight or flaps in a hurried or a distinctly down-pushing fashion, it isn't an eagle. If it is steady and unflapping, it may be an eagle.

Immature eagles usually fly solo. Adult eagles sometimes migrate in pairs, although the trailing bird may be miles and minutes behind. Vultures commonly migrate and forage in small flocks. In most places, a *flock* of big black birds is synonymous with *vultures.*

A distant Rough-legged Hawk (light or dark form) can be mistaken for a Golden Eagle, but the shapes of the dihedrals differ. In a Rough-legged, the uplifting begins at the shoulder and levels out toward the wrist. A Golden Eagle dihedral starts at the shoulder, flattens along the arm, and then lifts along the hand. Rough-legged, too, is known to rock in flight, a dead giveaway that the bird is not an eagle.

A morph Ferruginous Hawk, with its two white wing patches on the upperwing and its pale tail, most resembles an immature Golden Eagle — but just watch for a while. If the bird hovers or kites or rocks in flight, it isn't an eagle.

A high-flying Osprey may be mistaken for a Bald Eagle because of its large size, and because when an Osprey is in a full soar, the down-crooked wing is not as apparent. But Ospreys have a diminutive head; in fact, at the distances that Osprey and Bald Eagle are likely to be confused, Osprey seems all wing and tail, showing no head at all. Advanced immature Bald Eagles, showing lavish amounts of white mixed with dark patterning on the wings and body (like Ospreys), will have a large dark head that is visible at a distance. At tremendous distances (or very high overhead), adult Bald Eagles show no head *and* no tail. The birds look like a flying dash (—) in the sky.

On the other hand, a Bald Eagle approaching an observer head-on can easily be mistaken for an Osprey. Bald Eagle may glide with down-drooped wings that suggest Osprey, *particularly on days with very light winds.* Subadult eagles in transition between advanced immature and adult plumage can also have fairly white underparts, a white head, *and* an Osprey-like dark stripe through the eye. The best defense is simple mindfulness. An eagle may glide with down-drooped wings, but the wing configuration falls short of the exaggerated gull-winged shape common in Osprey.

Immature Golden Eagle (right) with immature Bald Eagle (left), soaring and turning away. Note slight dihedral of Golden. Compare shapes of both upper and lower wingtips. Bald is more angular, pointed; Golden is rounded, buteolike. Also compare tail length and shape, head size, and wing pattern.

Another way to determine whether a bird is an eagle is to watch and listen to the reaction of birds around you. If crows begin to get uncommonly raucous, or a perched Red-tailed begins screaming, it might be an eagle. If flocks of gulls, ducks, or herons and egrets suddenly take wing, it is very probably an eagle. If you do have bellwether flocks in the vicinity and they abruptly take wing, *start looking for an eagle.* Flocks sometimes flush for large buteos, but they seem to know the difference between, and act differently toward, vultures, Ospreys, and eagles. Unless very close, vultures and Ospreys are ignored. Eagles cause pandemonium.

When you've evaluated all possibilities and the bird is undeniably an eagle, the difficult part begins. Distinguishing eagles from eagles is far more difficult than separating eagles from birds of lesser lineage.

With their gleaming white wing patches, gleaming white tail, and absence of white on other parts of the body, boldly patterned juvenile Golden Eagles are easy to recognize. In fact, just the presence of gleaming white on a distant eagle (providing it's not on both the head and tail) means Golden Eagle. The white in the wings and tail of juvenile Bald Eagle is dingy, dirty white. The lavish amounts of white spattered all over the body of more advanced immature Bald Eagles (turning distant birds tawny or latte-colored) eliminates Golden Eagle from consideration.

A uniformly dark eagle that lacks any trace of white (except a modest amount of gray-white at the base of the tail) is almost certainly an adult Golden Eagle — but check to see whether the underwings are gray with dark borders and check to see whether the silhouette is buteolike.

Tail pattern is not a dependable way of distinguishing immature Bald and Golden eagles. Both immature eagles can have the same pattern, a mostly white tail with a well-defined dark band at the tip. In addition, adult female Goldens can show considerable amounts of white at the base of the tail (reminiscent of many first-year

Bald Eagles), and subadult Bald Eagles can have a bright white tail that still shows a narrow, dark tip (recalling juvenile Golden Eagles).

Juvenile Golden Eagles *do* have whiter tails that fairly gleam at a distance (tails of immature Bald Eagles are dirty white or infused with brown). Juvenile Golden Eagles also tend to have a more contrasting and more sharply defined tail band. These differences are useful but not definitive.

The location and distribution of white in the wings in immature eagles is, however, very useful for identification purposes. A juvenile Golden Eagle has white on the base of the flight feathers (about two-thirds of the way out from the body) that usually takes the shape of a large white patch, visible on the underside of the wing and sometimes the upper surface, too. In some immature Goldens, the white may be very restricted, limited to the base of the primaries and secondaries, forming a narrow broken white line running along the edge of the underwing linings. *On juvenile Bald Eagle the entire underwing lining, the leading edge of the wing, is dirty white.* Showing whitest of all on juvenile Balds will be the wingpits, appearing as white oval patches at the base of each wing.

Two- and three-year-old Bald Eagles, those showing lavish amounts of buff or dirty white plumage all over the body, can hardly be mistaken for immature Golden Eagles, which are either classically marked (juveniles) or mostly uniformly dark. Some second-year Golden Eagles may show some touches of white on their wings and body, but it is spare, not lavish. The body of Golden Eagles always appears dark, not pied or latte-colored.

In poor light, or at distances that make identification of an all-dark eagle uncertain (that is, distances where an adult Golden Eagle might be confused with a very dark, unmarked, immature Bald Eagle), the relative length of the head and tail is the most useful field mark. A bird with a tail that is about as long as its head or no more than one and a half head lengths is a Bald Eagle. A bird with a tail that seems at least twice as long as its head is a Golden Eagle.

Crested Caracara

A Chimera in Big Black Bird Clothing

SPECIES
Caracara cheriway

It is an anomalous bird — ranked among falcons, forged in the likeness of an eagle, feeding with vultures. Small wonder we puzzled over where to fit Crested Caracara into this book. With the tail of a young Golden Eagle, the wing pattern of a Black Vulture, and the overall head and shape of a Bald Eagle (with the crown of a Black-crowned Night-Heron thrown in for good measure), this bird looks as if it has been constructed out of spare parts — a chimera in raptor clothing.

Genetic allegiance notwithstanding, it made no sense to include North America's only caracara in the chapter on falcons. Distinguishing Prairie Falcon and Peregrine is a challenge worth addressing in this book; distinguishing Crested Caracara from American Kestrel is not.

A better case could be made for grouping the caracara among eagles and vultures, birds with which it associates and with which, because of its size, plumage, and shape, it might be confused. But while those other big black birds range widely across North America, the Crested Caracara is restricted. Over most of the United States and Canada, when confronted by the challenge of identifying a distant big black bird, the caracara is not a factor. Adding this species to the chapter on eagles and vultures would not serve the principles of the book.

While Crested Caracara's New World range is huge — from the southern United States all the way to Tierra del Fuego — in the United States, it is found only in the Florida prairie region, southwestern Arizona, and southern and coastal sections of Texas, ranging east to extreme southwestern Louisiana. In Florida, the bird is listed as "threatened," but the population appears stable; in Arizona, it is uncommon. Only in Texas is the caracara fairly common.

LEFT: An adult Crested Caracara in low direct flight over a Florida prairie. The loping, rowing wing beat hints at the Crested Caracara's falcon affinity, despite the unfalconlike wing shape. KK

Why, then, not include caracara in the chapter dealing with Florida specialties? Because while found in Florida, the bird's range is not limited to that state. Border buteos then? Chimera it is; buteo (even parabuteo) it is not. Regional specialties? A possibility . . . but while it is geographically restricted, the bird seems too widely distributed and too common to fit comfortably in a chapter featuring such rare and restricted species as Hook-billed Kite, California Condor, and Aplomado Falcon.

Once again, the anomalous nature of the bird distinguishes it. So we chose to honor caracara's distinctiveness by placing it in its own chapter, but also placing that chapter in comparative proximity to the big black birds, those species caracara seems most akin to.

If you are not hawk watching in an area that Crested Caracara frequents, be confident that there are only four big black birds in the deck of possibilities to choose from: Turkey Vulture, Black Vulture, Bald Eagle, and Golden Eagle. But when you cross that line and find yourself in Florida, Texas, or Arizona, be aware you are now playing with a deck of five — four big black birds and one wild card: Crested Caracara.

In Florida, caracara-sighting aspirants should search the cattle country north of Lake Okeechobee, where the caracara shares the Florida prairies with Sandhill Cranes and Burrowing Owls. In Texas, the bird is fairly widespread, but Aransas National Wildlife Refuge and Laguna Atascosa National Wildlife Refuge host particularly large populations.

Individuals have occurred outside their normal range — in California, New Mexico, Louisiana, Oklahoma, Mississippi, and elsewhere — but the bird is nonmigratory and shows only a slight tendency toward vagrancy.

As a raptor, the caracara is more opportunist than predator but as much predator as scavenger. Insects, worms, turtle eggs, and small (and occasionally large) vertebrates are part of its very eclectic diet. It forages early and in late afternoon, often touring highways for fresh roadkill. It follows tractors and cattle on the ground, snapping up any small creatures that get flushed. Like White-tailed Hawk, it is attracted to wildfires and burns, where it hunts prey dislodged by the flames.

But the caracara is a scavenger, too — a pseudo-vulture. It is usually dominant at carcasses, driving off one or several Black and Turkey vultures. But if a single caracara must compete with a large number of vultures, adult (and especially juvenile) caracaras often defer and wait for leftovers.

Shared scavenging habits make vultures excellent bellwethers for caracara. Vultures soaring over a kill, or perched or standing in an open field, suggest that a caracara is already there. If not — and if you're in caracara country — wait. Caracaras look for vultures, too. Sooner or later, one is likely to turn up — the consummate opportunist in search of an opportune meal.

Also, not only are caracaras opportunistic feeders (and pirates), they are gluttons, stuffing their crops. Not only does a well-fed caracara's crop protrude, it fairly glows! The unfeathered crop is bright pink to bright orange, bulges like a goiter, and is, as you might expect, visible from a great distance.

One final thought: To find caracaras, generally look low. The birds are more likely to be on the ground or on a perch than in the air. Their perches are commonly low — fence posts are perfect.

One final qualifier: Caracaras can, and do, soar well.

Crested Caracara, underside. Juvenile gliding **(A)**. Adult soaring **(B)**. Wings long and very straight edged, even width, black with pale whitish outer primaries; tail long and white with black tip; head long, obvious, whitish with dark face and cap. Long legs are conspicuous. Juvenile similar to adult but slightly paler, brownish; tail band pale gray-brown; breast streaked rather than barred.

Crested Caracara, upperside. Juvenile gliding **(A)**. Adult gliding **(B)**. Distinctive, dark with white "ends." Juvenile shows conspicuous pale brown panel on upperwing coverts, light brown tail band.

IDENTIFICATION

A roadside caracara perched on the ground or a fence post hardly presents an identification challenge. Distant or high-flying caracaras are another thing.

Hardly lacking for field marks, Crested Caracara brandishes a mix of traits borrowed from other big black birds. It is large and sturdy, averaging about the size of a Red-tailed Hawk. Crested Caracara is big bodied, big headed (unlike vultures), with a thick neck and a large, heavy bill. Legs are long and yellow. The bright orange face of an adult is visible at a considerable distance on a well-lit flying bird.

The body and wings are brownish black in adults. Juveniles are similar but not so black. The neck, upper chest, and upper back appear white on adults, buffy on juveniles and second-year birds. A close view will reveal fine black vermiculation on the adult breast and diffuse brown streaking on younger birds. At any age, the bold black cap is very conspicuous, particularly when raised in a crest if birds are alert or agitated.

IN FLIGHT. Wings and tail are the distinctive features of a Crested Caracara. Above and below — and contrasting with the overall dark wings and body — birds brandish a bright white tail tipped with a wide black terminal band. The tail is long and projects noticeably.

The dark wings are tipped with distinct white panels, visible at great distances. At close range, fine black barring is visible on the otherwise white wingtips, but fine points of plumage hardly count for anything where identification of caracara is concerned.

White neck and face, white tail, white wingtips, black wings and body: in essence and in sum, a Crested Caracara looks like a white-tipped plus sign (+) in the sky.

As with other falcons, the arm is short, the hand long. In a soar or a glide, the planklike wings are held at a stiff right angle to the body. The leading and trailing edges of the wings are parallel, but the hand is angled back enough to make the wrist project slightly.

Viewed head-on, gliding and soaring caracaras show a gentle down-curve to the wing that is akin to, but not as exaggerated as, the cupped wings of Great Blue Heron or Osprey. Viewed wing-on, the head and bill project prominently — as much as Bald Eagle's. Unlike the eagle, caracara flies with the head drooped. It gives the bird a hunchbacked appearance.

Although caracara can and does soar at great altitude for extended periods, this is the exception rather than the norm. For the most part, the flights are low, active, and direct — a lengthy series of steady, fairly hurried wingbeats

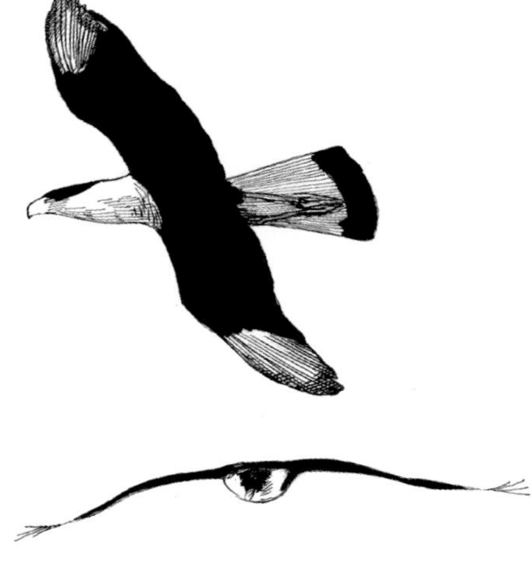

Crested Caracara adult gliding. Side view and head-on.

punctuated by a short, stiff-winged, droop-winged glide. Even while flapping, the wings retain the cupped configuration.

Wingbeats are shallow, fluid, fairly fast, and steady — very unvulturelike. There is a hurried quality to them. Equal emphasis is given to the upstroke and downstroke, a quality the bird shares with other falcons (in fact, and despite radically different wing shapes, the bird's flight is reminiscent of a Peregrine — up to a point). The wingbeats also have a rowing quality that recalls a Pileated Woodpecker in flight.

The flight is direct and steady, with none of the tippiness of Turkey Vulture or hurried flap of Black Vulture. Overall, the GISS of the bird is robust and powerful — a large, broadly proportioned bird that puts a lot of effort into each wing stroke but shrugs off the hardship.

When all the anomalous parts and qualities harmonize to form a caracara in flight, the bird appears much more eaglelike than anything. Distinguishing Crested Caracara really comes down to that.

1. A juvenile Crested Caracara banking in a soar, from above. It looks like a bird made out of spare parts. Juveniles are mostly brown, adults black. The basic pattern — black-capped white head, white patches on wingtips, dark-banded white tail — holds true no matter what the sex or age. Texas, December NH

2. An adult Crested Caracara in a glide, showing white at every compass point: head, wingtips, tail. Adults have blackish plumage and dark speckling, not streaking, on the hindneck and breast. Texas, December NH

3. An adult Crested Caracara in powered flight. Birds fly with a strong, rowing wingbeat and often fly and maneuver with what seems like reckless speed for such a large bird. Then again, caracara *is* a falcon! Texas, December NH

4. Two adult Crested Caracaras — left bird banking in a full soar (note the planklike wings set at an almost perfect right angle to the body), right bird in a fast glide with wings gently drooped below the body. Veracruz, December CS

5. Juvenile (left) and adult (center) Crested Caracaras pursuing a juvenile White-tailed Hawk carrying prey. Piracy is part of the caracara's game book. Smart money says this White-tailed Hawk is going hunting again. Texas, December NH

PUTTING IT TOGETHER: Telling Caracaras from Other Big Black Birds

Yes, Crested Caracara is a chimera with the head shape and color of a Bald Eagle, the tail of a young Golden Eagle, and the wings of a Black Vulture.

But a young Golden Eagle would show white at the base of the flight feathers, never the white panel at the tip of the wing as caracara does. The Black Vulture, which *does* show white primary panels, would never show any white on the tail. A juvenile or subadult Bald Eagle would always show some white on the underwing linings (and often on other parts of the wings and body). On caracara the white is strictly segregated. Head. Tail. Wing*tips.*

None of these potential candidates for confusion flies hunchbacked with a rowing wingbeat, and none of these birds shows white at every projecting point: head, tail, wingtips.

Based on shape alone, with no plumage points visible, a distant caracara most closely resembles an Osprey or an adult Bald Eagle. Osprey soars with cupped wings and wingtips swept slightly back. But at almost any distance, a soaring Osprey will eventually display its white underparts (and unlike caracaras, Ospreys have a no-headed appearance). While Bald Eagles may glide with wings drooped, they never appear hunchbacked.

The only other bird that might be confused with Crested Caracara is a raven. Ravens are overall dark and similarly proportioned — projecting head, projecting tail, wings that droop, and wingtips that curl back. Even the rowing flight is similar in manner if not in cadence (the flight of ravens is languid, not hurried).

While it is unlikely that a caracara will be mistaken for a raven (which has no white on the body, wings, or tail), a distant caracara might easily be *dismissed* as "just a raven." Be advised, and be sure to study all distant ravens.

Osprey

The Fish Hawk

SPECIES
Pandion haliaetus

The Osprey is a large, strikingly handsome raptor with its own evolutionary specialty — diving for fish.

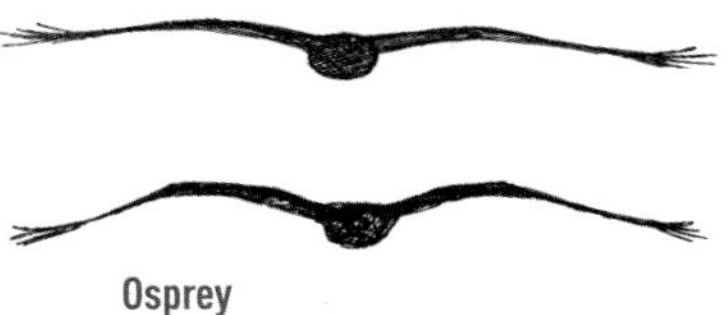

Osprey soars with wings nearly flat, with just a slight downward bow. When it is gliding, the wings are strongly arched, similar to the wing position of a large gull.

Bald Eagles capture fish that swim close to the surface by flying low and casting their talons into the water. But only the Osprey *dives,* entering the water talons-first, often immersing itself completely in pursuit of piscatorial prey.

A hunting Osprey will fly, flapping slowly, heading into the wind, to a likely patch of water and hover 20 to 100 feet above the surface with an angler's patience until it sights prey. When all seems right, the bird folds and dives, headfirst, with wings swept back behind the tail. The bird adjusts the angle of the dive to compensate for the refracted image — a neat trick and an essential mechanism for a fish-catching bird. Just before hitting the water, the Osprey throws its feet forward in an eight-cornered net of talons.

More than half the time, an adult Osprey will emerge with fish in tow. It will lift off the water with a single, powerful wingbeat (or a series of wingbeats if the fish is unusually large). After a shivering shake to shed water droplets from its oil-saturated plumage, the bird heads for a perch to savor its catch, providing it can ward off the pirating efforts of gulls, Bald Eagles, frigatebirds, and other Ospreys.

In migration, an Osprey may catch and carry, or "pack a lunch," as some call it. Ospreys are often seen moving past hawk-watch sites with fish in their talons, far from any known fishing hole. On one occasion at the Montclair Hawk Watch in

LEFT: An adult Osprey. The "Fish Hawk" is a unique specialist and uniquely patterned. KK

New Jersey, an Osprey was reportedly seen carrying an orange — unlikely prey for a fish-eating bird, especially in a temperate zone. Later reflection led observers to conclude that the bird was most likely carrying a large goldfish appropriated from someone's ornamental pond.

Osprey may infrequently and perhaps inadvertently take prey other than fish. A snake swimming across the surface of a lake is fair game, and muskrat skulls have been found in Osprey nests, although perhaps the latter simply formed part of the flotsam with which Ospreys adorn their impressive stick nests. Nest-building materials may include an array of natural and unnatural items, from seaweed to conch shells to the desiccated remains of Great Blue Herons to the nylon webbing from beach chairs and Styrofoam coolers.

The nests, made mostly of sticks and driftwood, are impressive structures. Pairs add material to them year after year until a storm destroys the structure or a supporting base collapses under the sheer weight of material. Mostly building among the topmost limbs of dead trees, Ospreys also build nests atop manmade structures such as channel markers, utility poles, duck blinds, and grounded boats. They also readily adopt artificial nest platforms erected specifically for their use. Prime nesting locations, situated on coasts or lakeshores and forest or wooded islands near marshes, offer unobstructed views, good security, and nearby sentry perches.

The bird is usually a solitary nester, with pairs occupying and defending their own private, waterfront territory. In some places, though, Ospreys form loose colonies.

Osprey, as a single species, occurs over much of the globe but most commonly in the Northern Hemisphere. In North America, the subspecies *Pandion haliaetus carolinensis* ranges widely as a breeder across much of the northland, from Alaska to Newfoundland and south into central California, the Rocky Mountain West, the Great Lakes, and along the entire length of the Atlantic Coast as well as the Gulf Coast from Louisiana to Florida. Breeding density varies greatly across the bird's range and is, needless to say, very much related to proximity to fish-bearing bodies of water.

Birds winter in California, southern and eastern Texas, and coastally from Louisiana to North Carolina (including almost all of Florida). Most Ospreys, however, retreat farther south to the Caribbean and Central and South America, the only continent (except Antarctica) where Ospreys do not breed. Immature Ospreys remain on their winter range the summer after they fledge, returning as breeders the second spring of their lives.

MIGRATION

This is a long-distance migrant. With some birds wintering as far south as Argentina and central Chile, Ospreys rival Swainson's Hawks and Peregrine Falcons as North America's long-distance champions. In spring, birds leave South America in February and reach breeding territory in the Middle Atlantic States as early as the first week in March. The peak of the migration occurs from April into early May. Birds begin heading south during the middle of August. Peak fall migration occurs in late September and October. Other than Florida residents and a few stragglers elsewhere, by November, Ospreys have largely departed from most of the United States and Canada.

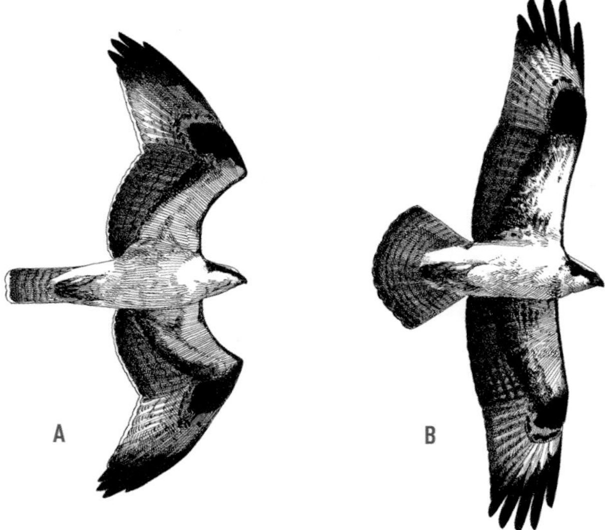

Osprey, underside. Juvenile **(A)**. Adult **(B)**. A distinctive bird, with long crooked wings. Little plumage variation. All have white body, alternating dark and light areas on underwing including obvious black carpal patch; white head with black eye stripe. Juvenile differs from adult in having white tips on secondaries and tail, more translucent inner primaries.

Osprey, upperside. Great Black-backed Gull **(A)**. Adult Osprey **(B)**. Juvenile Osprey **(C)**. Osprey shape is most similar to gulls', and plumage is very similar to adult Great Black-backed Gull's **(A)**, but distinguished by blackish rump and tail, visible at tremendous distances. Adult **(B)** is uniformly blackish brown above, browner on coverts when feathers are worn. Juvenile **(C)** in fresh plumage has all feathers tipped with white.

Significant numbers of migrating Ospreys are recorded at interior-ridge hawk-watch sites in fall. Spring migration along interior ridges is likewise good, often rivaling autumn totals. Very impressive numbers of Ospreys also occur at coastal sites in fall and at Great Lakes sites in spring and fall.

Ospreys fly solo, in pairs, and, in places and at times, in small groups. It is productive when an observer has spotted an Osprey to scan ahead of the bird and then behind to pick up leading or trailing birds.

Ospreys migrate later in the day than most other raptors, sometimes moving past hawk-watch sites as late as observers are able to see birds in the growing darkness. They are much less inclined to rise early, preferring to wait until thermals start forming near midmorning.

Ospreys are graceful fliers that seem capable of finding lift when other birds cannot. Soaring in a kettle of Broad-winged Hawks, they easily outclimb the smaller buteos. This strong flier also resorts to active flight, particularly over open water.

IDENTIFICATION

An Osprey is a big raptor with distinctive blackish-brown-and-white-patterned plumage and an equally distinctive shape. It is very probably the easiest North American raptor to identify. Just *look* at the illustration. It pretty much says it all. Only the Bald Eagle or Great Black-backed Gull is a candidate for confusion, with Swallow-tailed Kite an outside possibility.

Adult and juvenile Ospreys have similar plumage. Adults are dark brown, almost black above, except for a boldly patterned head, whose most prominent feature is a narrow black mask set over the bird's eyes and face. Below, the body and underwing linings are white, contrasting noticeably with the darker trailing edge of the wing. Large carpal patches, which are blacker than the outer flight feathers, are very conspicuous.

Juveniles are slightly lighter, browner above, but otherwise much like adults. At close range, juveniles show pale feather edges on the backs and upperwings, which makes more distant juveniles appear scaly and very distant birds appear more brownish than blackish backed, which is typical of adults.

Both males and females can have dark streaking on the upper breast—a "necklace"—although the necklace on females is, on average, darker and heavier.

IN FLIGHT. Ospreys have a small head, medium-length tail, and exceedingly long wings. The wings seem to dominate the bird; in fact, when birds are viewed flying high overhead, the head simply disappears.

In a full soar, an Osprey's wings are uncommonly long and narrow. The arm is long and curved forward. The hands are long, narrow, and angled back. The

An Osprey carrying a fish.

effect is very gull-like *except* that on Osprey, the outer flight feathers are deeply slotted. (Gulls are like Peregrines — their wings simply come to a point.)

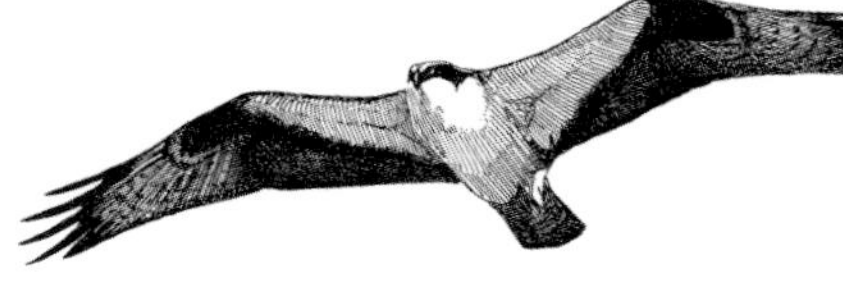
An Osprey gliding.

In a glide, this gull-like wing configuration becomes acute. The bird's arm *juts* forward and seems to shorten (the bird's wrists actually reach the tip of the bill). The arm angles sharply back, tapers, and seems to lengthen, dominating the wing. With its deeply slotted flight feathers (still visible when the bird is gliding), the bird resembles a long-winged scarecrow with its arms thrown over the crossbar of a supporting post. Many observers liken the shape to a flying M.

Seen head-on, the bird holds its wings in an exaggerated uplifted bow completely above the horizontal axis (a head-on M). The bird glides as if it were riding with all its weight placed on the palms of its hands.

A fish's-eye view of a juvenile osprey. CS

In brief, gliding or soaring, the Osprey silhouette is distinctive. Or, in the words of hawk watcher Harry E. LeGrand, "It's a hawk that looks like a gull, but it's not."

No description says it better.

From almost any angle, the bird appears small headed, a trait accentuated by the length of the wing and the wrists that thrust forward in a glide. The tail is of medium length but broad. When the bird is flying high overhead or when seen from the side, the head fairly disappears. A distant Osprey appears all wings and tail.

The wingbeat of an Osprey is stiff, almost arthritic. The motion seems centered at the wrists, and the hands don't contribute to the wing action — in fact, when flapping, the hands just seem to go along for the ride or play catch-up to the rest of wing.

When the bird is flapping heavily, the body (particularly the head) bobs up and down in counterpoint to the stroke of the wing. The stiffness, the rhythm of the wingbeat, and the bobbing motion suggest the movement of a figure on crutches moving hurriedly. During migration the Osprey uses thermals and updrafts extensively but may engage in more active flight, particularly when making broad, open-water crossings.

Hovering birds raise their wings high over the body, unlike Red-tailed Hawk, whose hovering flight is mostly downstroke.

1. A juvenile Osprey hovering. Since it's a fish-eating raptor, you can bet what it is hovering over is open water. Florida, October KK

2. An adult Osprey in active flight, carrying a partially eaten menhaden. Wingbeats are deep and choppy, somewhat anxious or hurried. Often the body bobs up and down, too light for the bird's powerful wingbeats. New Jersey, July KK

3. An adult Osprey gliding and banking. The size, shape, and plumage are so distinctive that any explanation relating to a bird seen this close and this well is superfluous. However, the Osprey's classic M-shaped configuration is well shown here. New Jersey, July CS

4. A juvenile Osprey in a full soar. Even with wings fully fanned, something of the M-shaped wing configuration endures. Things to note in this image include the very clean trailing edge to the wing and the buffy underwing, characteristic of a juvenile, as opposed to the clean white underwing of adults (see #3). Cape May, October CS

5. An adult Osprey gliding. The classic M- or gull-shaped configuration of an Osprey's wings is apparent head-on, too. Bald Eagles, seen head-on, occasionally droop their wings enough to be confused with an Osprey, but they never show such a strongly M-shaped wing. Utah, May JL

6. An Osprey gliding high overhead, going away. The classic pattern and shape are identifiable at great distances. Cape May, October CS

7. The classic view of an Osprey migrating high overhead. At higher altitudes, the head gets lost in the projecting jut of the wings, making the bird look headless, like just an M with a tail. Cape May, October CS

8. A gliding adult Osprey, seen from above. The uniformly blackish brown upperside is distinctive. Juveniles would have pale edges to the back feathers, making the bird appear scaly. Not apparent here, but the tail of backlit birds often appears reddish, and confusion with Red-tailed Hawk is remotely possible. Cape May, September JL

PUTTING IT TOGETHER: Telling Ospreys from the Rest

The Osprey seems unlikely ever to be confused with another bird, but it sometimes is. Distance diminishes dissimilarity, and at a certain point even a bird as distinctive as an Osprey challenges the observer's skills.

The bird's resemblance to a gull has been mentioned and cannot be overstated. Ospreys have also been confused with distant soaring eagles (and vice versa). Then, too, there have been occasions when inexperienced hawk watchers have taken an Osprey down to its elemental field marks — light leading edge to the wing, dark trailing edge, dark back, bib (necklace), long slim wings — and come up with "Swainson's Hawk!"

And more than one Osprey has been called Red-tailed Hawk by an incautious observer because, when an Osprey is backlit, the sun transfusing through the tail makes the tail appear distinctly rufous — something field guides don't mention, and why would they? Who is going to mistake an Osprey for a Red-tailed Hawk?

At anything approaching a reasonable distance, an Osprey should be instantly recognizable. Even at distances, altitudes, and light conditions in which even plumage patterns as distinctive as those on an Osprey are useless, the distinctive, gull-shaped wing configuration should eliminate all other birds except Bald Eagle, large gull, Swallow-tailed Kite, and perhaps a gliding light-morph Swainson's Hawk.

Bald Eagles do glide or soar with their wings slightly drooped, recalling Osprey. But with eagles, the wings start flat and droop toward the tip. An Osprey's wings angle up from the shoulder, then angle down at the wrist. It's not a gentle curve; it's a shallow M.

Bald Eagles also have massive, projecting heads. The Osprey's head is small and inconspicuous.

Sometimes an observer can look too closely and be confused, not comforted, by plumage. The head pattern of immature Bald Eagles in advanced stages of maturation can mimic the masked look of an Osprey. If you are close enough to see a dark eye line, you are close enough to see that an immature Bald Eagle has underparts that look spattered by white paint and mangled by molt. While some whiter-bellied Bald Eagles are extensively bright white below, they don't have the clean, crisp patterning of an Osprey.

Distant gulls, particularly adult Great Black-backed Gulls, can be mistaken for Ospreys, although more often, when numbers of gulls are found near a hawk watch, Ospreys are more likely to be overlooked than misidentified. A distant Osprey will appear to have a small head or no head; the bird is just wings and tail. Gulls have large conspicuous heads that extend well beyond the wings. Furthermore, Ospreys have broad wings that may recall the long, slim tapering of a gull's wings, but they do not replicate it. An Osprey's wings are broader than a gull's, and the wingtips are slotted when soaring and ragged tipped when gliding. Soaring or gliding, a gull's wings are knife edged and pointed; individual flight feathers are indistinguishable.

The wingbeat of the larger species of gulls is methodical, languid, and shallow, with overtones of effortlessness. The wingbeat of an Osprey is stiff, labored, pumping, suggesting real effort. On average, gulls flap a good deal more than do Ospreys.

At many, if not most, hawk-watch sites, gulls and Ospreys use different flight patterns. Gulls are not attracted to updrafts as much as Ospreys are, and/or their daily foraging pattern ferries them across, not along, the ridge. Ospreys are compulsive updraft users and will follow the contour of a ridge.

One final observation: Gulls, for all their aerial finesse, hover for short duration with obvious effort and little finesse. Ospreys, on the other hand, can hold steady in the sky and hover for as long as it suits them — sometimes minutes at a time.

Southwestern Buteos and Kin

At Home Where North Meets South

SPECIES
Gray Hawk, *Buteo nitidus*
Common Black-Hawk, *Buteogallus anthracinus*
Zone-tailed Hawk, *Buteo albonotatus*
White-tailed Hawk, *Buteo albicaudatus*
Harris's Hawk, *Parabuteo unicinctus*

Sometime after becoming enamored of raptors, you will be drawn to the southwestern deserts, grasslands, and mountains and a wonderful group of hawks found only there. All are distinctively plumaged, and most make a fashion statement with a minimum of colors — black, white, and gray (though White-tailed and Harris's Hawks add a tasteful touch of rust).

For raptor spotters and for identification purposes, all the raptors included in this chapter are birds of the same ilk — broad-winged, broad-tailed, soaring birds at home in the skies that stretch from the humid haze of the Texas coast to the halcyon blue of the southwestern deserts.

While these southwestern specialties are peripheral species in the United States, all are more common in Mexico and Central and South America. Their celebrity status among North American birders has to do with their restricted United States range and, of course, their status as raptors.

MIGRATION

While sedentary over most of their range, Gray Hawk, Common Black-Hawk, and Zone-tailed Hawk mostly withdraw from their United States range in winter and are present mainly from March through September or early October. White-tailed Hawks are found in Texas year-round, yet numbers increase markedly in winter at

LEFT: An adult Harris's Hawk in a glide. The Harris's is one of the most colorful of all hawks — in both plumage and lifestyle. NH

Southwestern hawks: soaring juveniles of all species, backlit. White-tailed, Zone-tailed, and Gray hawks are typical buteos. Common Black-Hawk and Harris's Hawk are not in the genus *Buteo* but are closely related to it. Hook-billed Kite is superficially similar to buteos in some respects but unrelated. White-tailed Hawk **(A)**: wings long, pointed, and gracefully curved, pinched in at body, overall mottled dark gray-brown with translucent tail. Zone-tailed Hawk **(B)**: long even-width wings and fairly long tail, uniformly blackish on body with at most a few white speckles on breast. Juveniles show faint translucent windows on inner primaries. Gray Hawk **(C)**: small, long tailed, and stubby winged, all flight feathers pale and strikingly translucent.

Common Black-Hawk **(D)**: extremely broad winged, overall dark, heavily streaked but with obvious whitish bands in tail and translucent patch at base of inner primaries. Harris's Hawk **(E)**: broad, rounded wings and long tail; wings especially broad in secondaries; dark chest and underwing coverts contrast with white at base of tail and pale inner primaries. Hook-billed Kite **(F)**: wings extremely broad tipped, paddle-shaped, all primaries translucent. *Note: Species treatment of Hook-billed Kite is found in Chapter 12.*

An adult White-tailed Hawk soaring. The White-tailed is one of the great raptor rewards of the Texas Coast. NH

coastal, food-rich areas such as Laguna Atascosa National Wildlife Refuge. Harris's Hawks do not appear to migrate, although they exhibit some nomadic tendencies, movements that are presumably linked to rainfall and prey abundance.

All of these southwestern birds do evidence some limited to modest pattern of vagrancy, and Zone-tailed Hawk has been recorded well outside its expected range. Common Black-Hawks are recorded annually at Rio Grande Valley hawk-watch sites, such as Bentsen–Rio Grande Valley State Park and Santa Ana National Wildlife Refuge, and a northbound April movement of this species has recently been noted in Arizona along the Santa Cruz River.

In the "River of Raptors" flight over Veracruz, Mexico's famous migratory bottleneck, all five of the species under discussion are seen in small numbers among the huge streams of hawks heading south. Though many are presumed to be locals, a number are counted, as behavior suggests that the birds are engaged in a limited form of migration. From the standpoint of studying and identifying raptors, whether such behavior is labeled dispersal, migration, or a local movement hardly matters. The identification challenge remains a constant value.

Gray Hawk

Southeastern Arizona's San Pedro River flows north out of Mexico, forming a verdant corridor and a migratory pathway for birds. The cottonwoods and willows that lie along the 50-odd-mile stretch of river extending north of the Mexico–United States border constitute some of the most important riparian habitat in the West. Now controlled by the Bureau of Land Management, the 56,000-acre San Pedro Riparian National Conservation Area is home to many bird species whose range barely brushes the United States border. These include the Gray Hawk. Some 24 pairs, approximately 25 percent of the United States population, are found along the river corridor. Here, their plaintive whistles emanate from the streamside cottonwoods, and their gray-and-white silhouettes are etched against the skies.

Gray Hawk has one of the most restricted ranges of any of the southwestern specialists covered in this chapter (second only to the much more common White-tailed Hawk of coastal Texas). Its entire United States population may be fewer than 100 pairs, most of which are found in Arizona. A limited number of birds nest in extreme southwestern New Mexico as well as in the Rio Grande Valley of Texas (principally between Brownsville and Falcon Dam and in Big Bend National Park).

Feeding primarily on lizards, most Gray Hawks withdraw south in winter, leaving the United States in late September or early October and returning in March. A few individuals winter in the Rio Grande Valley.

During the breeding season, within its restricted United States range, the adult Gray Hawk should not pose a tough identification challenge. The problem may be more one of finding the birds, for despite their dashing flight and high-strung behavior, nesting Gray Hawks are as secretive as any other raptor that does not want to draw attention to its nest or young.

IDENTIFICATION. The Gray Hawk is a small buteo, similar in size and shape to a Broad-winged Hawk, the species with which it is most likely to be confused. It was once called the Mexican Goshawk thanks, in part, to its dashing, accipiter-like habits and flight, and in part because of its more-than-passing resemblance, in plumage, to an adult Northern Goshawk.

Above, an adult Gray Hawk shows a medium-gray back, generally lighter than the backs of adult Sharp-shinned Hawk or Cooper's Hawk. It can appear slate-colored in dim light but exceedingly pale in harsh desert sunlight. The gray back is a useful characteristic when searching among myriads of brown-backed Broad-winged Hawks in southern Texas.

Below, an adult Gray Hawk is pale, possessing white underwings and coverts, with fine gray barring across a pale belly and chest. These vermiculate markings merge and disappear at a distance, making underparts look pale gray or dirty white. The chest is distinctly darker than the belly. The tail is prominently marked with broad black-and-white tail bands that are particularly prominent when birds are backlit. Unless backlit (and like adult Broad-winged), it is common for only a single broad white tail band to show prominently.

Juvenile Gray Hawk is brown above, with a multibanded tail and a bold face

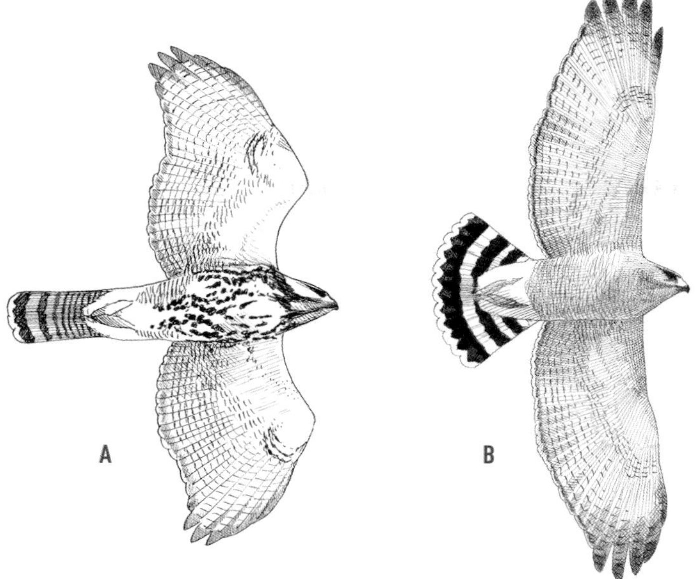

Gray Hawk, underside. Juvenile gliding **(A)**. Adult soaring **(B)**. More reminiscent of an accipiter than a buteo, especially juvenile, which has longer tail and wings than adult. Wings relatively short and broad, tail long. Note slight bulge on outer secondaries; wings slightly tapered but blunt tipped. Adult unmistakable, with pale gray plumage and boldly banded tail. Juvenile has heavy spotting on body and bold face pattern but relatively unmarked underwing coverts and flight feathers; dark trailing edge on wing is very faint.

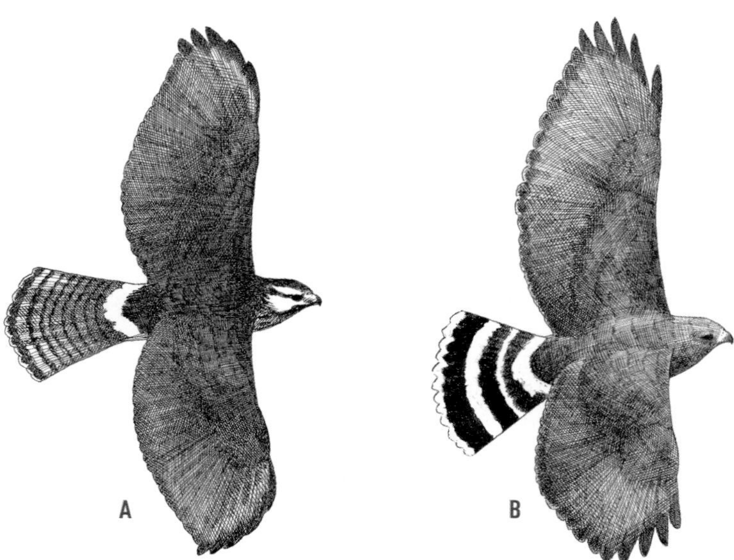

Gray Hawk, upperside. Juvenile soaring **(A)**. Adult soaring **(B)**. Fairly uniform plumage, gray on adult, brown on juvenile, except boldly barred tail of adult and bright white uppertail coverts of juvenile.

pattern featuring a dark eye line and malar stripe contrasting with a bright white supercilium and white cheeks. Juvenile Broad-winged shows a similar pattern but, owing to its dark cheeks, the face pattern on Broad-winged is distinctly less bold and contrasting.

Juvenile Gray Hawk in level flight. Distinguished from accipiters and small buteos by striking face pattern and clean white uppertail coverts.

The underwings are finely patterned but seem pale and unmarked at a distance. The white underbody is heavily marked with thick gray-brown streaks that break into dark spotted strands on the belly.

Overall, the face and underbody of juvenile Gray Hawks are more heavily patterned and show more contrast than those of juvenile Broad-wingeds; markings are darker, cold brown as opposed to the rufous brown on the latter.

Both adults and young show a prominent white U-shaped band at the base of the tail (dorsal side). Other species (including Swainson's Hawk and Broad-winged Hawk) also show white at the base of the tail, but not the well-defined U pattern found on Gray Hawk. While not always apparent under all field conditions (particularly on adults), the U-shaped band is nonetheless a clincher field mark.

IN FLIGHT. Overall outline resembles a blunter and blockier but longer-tailed Broad-winged Hawk or perhaps a Broad-winged with Northern Goshawk sympathies. The wing shape is, at once, a hybrid of several species and unique, combining the overall basic candle-flame-shaped configuration of Broad-winged, the small-handedness of Northern Goshawk, and a hint of the pinch-based, paddle-winged configuration of Harris's Hawk. The wings of adults are blunter and more S-shaped along the trailing edge than adult Broad-winged's (suggesting a broader-winged Northern Goshawk), while narrower-winged, clean-lined juveniles have wings that more closely approximate the classic candle-flame configuration of Broad-winged Hawk.

Adult Gray Hawks appear slightly longer tailed than Broad-wingeds, but juvenile Gray Hawks do have a long tail, distinctly longer in appearance than the tail of a juvenile Broad-winged Hawk, making juvenile Gray Hawks appear rangier and more accipiter-like.

Gray Hawk behavior is an aid to identification. Gray Hawks are more high-strung than most buteos, and their nimble maneuvers recall an accipiter. They will often dash about, under, and through the canopy. When soaring, they may stoop to perch rather than glide in the staid manner of most buteos. Gray Hawks often soar at low altitudes (treetop level), making tight, sharply banking turns while flapping frequently.

They are quite skilled at maneuvering, in part because of the short wings and longish tail. Leaving a perch, Gray Hawks seem to accelerate their wingbeats. The first and second are choppy, hesitant; then comes a rapid blur of wingbeats almost too fast to count, again very accipiter-like.

The flap is hurried and stiff in cadence and execution, recalling a female Cooper's Hawk. Unlike most buteos, which are content to glide into a soar, the impatient Gray Hawk flaps into a soar using a batty series of flaps—sometimes three . . . eight . . . five . . . there's no predicting. This species turns tightly in a thermal and gains height very quickly.

In fact, in every facet of flight, Gray Hawks do things just a little bit faster than your experience with other buteos tells you they can or should.

Glides may be protracted, and often the angle is acute. The bird seems anxious to trade altitude to gain speed and time (in fact, it will often interrupt or punctuate a glide with active, wing-pumping flight).

In a shallow glide, wings may appear flat. More often, the wings are bowed even when flapping. When the bird is soaring, the wings are flat, with just the wingtips turned up.

Gray Hawk is not difficult to separate from any of the other southwestern specialties in this chapter. The similarly sized Broad-winged Hawk is most similar to Gray Hawk. While the Broad-winged is a rare migrant where Gray Hawks occur in Arizona, in Texas Gray Hawks are occasionally found among hundreds (even thousands) of Broad-wingeds migrating through the Rio Grande Valley in spring and fall.

When gliding, Gray Hawks appear more angular, with wings more drawn in, than Broad-wingeds (which always show a gentle, curving, swept-back look to the wing). In a soar, the raised wingtips of Gray Hawks are readily distinguished from the down-angled wings of Broad-winged Hawks. The blunt, squared-off look of the wingtip recalls Red-shouldered Hawk.

In a soar, the Gray Hawk's tail is not generally fanned as much as a Broad-winged's. It is more like an accipiter's partially spread tail. The head of a Gray Hawk seems to project more than a Broad-winged's and appears narrower, but this difference is very subtle. A final point to note is the translucent flight feathers of adult Gray Hawks, more so than Broad-wingeds. When backlit, the hand, and often the entire trailing edge of Gray Hawk's wing, seems to glow with transfused light.

Juvenile Gray Hawks may look and act very much like Cooper's Hawks. The extremely long tail of juveniles can give it a shape very reminiscent of Cooper's Hawk, but the striking dark-and-white markings on the face of juvenile Gray Hawk easily distinguish it from the warm brown hooded face of juvenile Cooper's. It is when gliding that juvenile Gray Hawks are most accipiter-like (and observers are most likely to be fooled). When soaring, they are clearly buteos (long tail notwithstanding).

1. An adult Gray Hawk in full soar. Somewhat like a Broad-winged Hawk in black and white — but a blunter-winged Broad-winged that shows more white bands in the tail. Arizona, June NH

2. An adult Gray Hawk showing bulging secondaries and squared-off wingtips. Arizona, March NH

3. An adult Gray Hawk in a glide. One look at the broad, blunt wings tells you this is not a bird that migrates very far. The long tail, in combination with the bird's very nimble and active hunting behavior, imparts a very accipiter-like GISS to the "Mexican Goshawk," as it once was known. Arizona, June NH

4. A juvenile Gray Hawk showing characteristic pale, unmarked underwings, dark and contrasting dark chocolate dribbles on the face and underparts. The facial pattern, showing a dark-bordered white cheek, seems starkly contrasting. If you are thinking that this bird is shaped more like a Broad-winged Hawk than the adult, you're right! Arizona, May NH

5. A gliding juvenile Gray Hawk. Now you see how this bird might be mistaken for an accipiter. This darker individual in a glide shows short broad wings and an accipiter-long narrow tail. Texas, February TJ

6. The dorsal view of a banking juvenile Gray Hawk (same bird as #5), showing the white U at the base of the tail, as well as the classic contrasting white-cheeked face pattern. TJ

Common Black-Hawk

Seeing a Common Black-Hawk soaring over the cottonwood canyons it calls home is one of hawk watching's great rewards. In the United States, it is also one of hawk watching's greatest challenges. Rare and local, Common Black-Hawk is also somewhat sluggish and sedentary, given to sitting for long periods on strategic limbs, watching for prey. Many observers make several attempts to find the bird before meeting with success, and most sightings are of perched, not flying, birds.

Not a true buteo (it's in the genus *Buteogallus*), Common Black-Hawk is nevertheless buteolike enough to be called buteo in the hawk-watching arena. In the United States, Common Black-Hawks nest in the Guadalupe and Davis mountains of Texas, New Mexico, and southwestern Utah. It has been recorded as a vagrant in California, southern Nevada, southern Colorado, and the Texas Panhandle. But, as with the Gray Hawk, Arizona is the species' North American stronghold. South of the United States–Mexico border, the bird is much more common and widespread, found along the Pacific, Gulf, and Atlantic coasts south to Colombia and Venezuela.

It is listed as threatened in Texas, endangered in New Mexico, and a species of concern in Arizona. There are an estimated maximum 250 pairs in the United States (most in Arizona) and, while it appears stable, owing to its limited range and habitat requirements, the United States population is vulnerable.

An obligate riparian species, it frequents mature, riparian woodlands along permanent streams, searching for mostly aquatic-based prey: crawfish, fish, and amphibians (as well as land crabs, lizards, snakes, small mammals, and nestling birds).

Over most of its range, this species is a permanent resident. In the United States, it is migratory, withdrawing south for the winter — leaving in October and returning in March. It is recorded as a migrant at Rio Grande Valley hawk watches, and a modest spring migration has been noted in southeastern Arizona.

Several hints for finding Common Black-Hawks, particularly if you want to enjoy them in flight: Use bird-finding guides, but go early in spring. They are very active during courtship (late March to mid-April), but during much of the summer they are more sedentary and soar little. In early spring, they soar high, often, and for prolonged periods, sometimes up to an hour. Also, during the heat of midday, birds sometimes bathe, then soar high over the canopy (perhaps to cool off). But if you plan to stake out a watering hole in Arizona in summer, in the noonday sun, plan to immerse yourself as well.

IDENTIFICATION. If finding Common Black-Hawks is the hard part, identifying them is the easy part, for they are one of our most distinctive hawks — a "buteo" with a shape reminiscent of Black Vulture.

Common Black-Hawk is big and stout, with the wingspan of a Red-tailed Hawk, but much more bulk. Adults are coal black — excellent camouflage for a species that strives to disappear into the deep shadows of streamside cottonwoods. *Note: Soaring, well-lighted birds may show a warm reddish cast in the primaries and secondaries (especially prominent on worn adults).* The long, bright yellow legs and

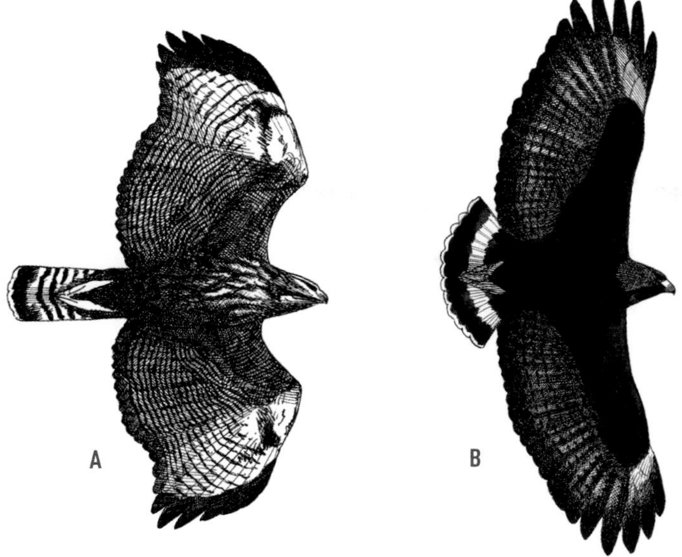

Common Black-Hawk, underside. Juvenile gliding **(A)**. Adult soaring **(B)**. Shape distinctive, adult has shorter tail and broader wings than juvenile such that trailing edge of wings overlaps sides of fanned tail and produces a nearly continuous line from wingtip to wingtip. Juvenile less pronounced but still has unusually broad wings, particularly the secondaries. Gliding shape distinctive, hunched and blocky. Adult uniformly blackish on body and underwing coverts but flight feathers grayish with broad dark tips; also white tail band and small whitish patch at base of outer primaries. Juvenile dark, heavily streaked on body, but primaries and primary coverts quite pale; tail white with many wavy dark bands.

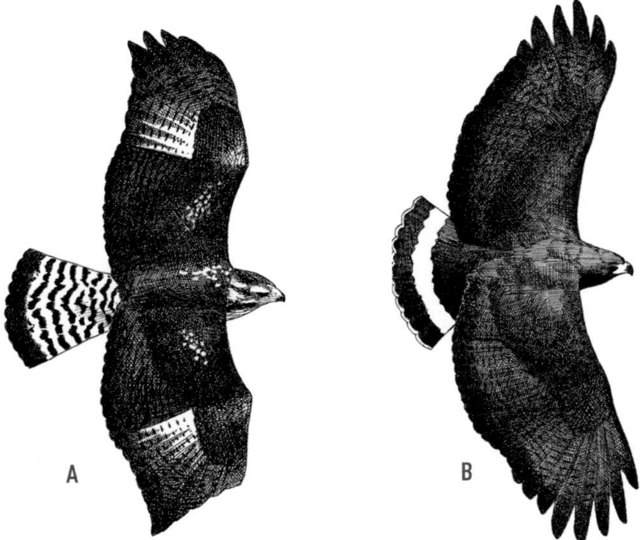

Common Black-Hawk, upperside. Juvenile gliding **(A)**. Adult soaring **(B)**. Adult dark blackish except tinged brown on flight feathers with faint dark trailing edge evident; single white tail band always prominent, unlike Zone-tailed Hawk, which shows white above only when tail is fully fanned. Juvenile dark brown with unique buffy patches at base of inner primaries; tail white with wavy black bands.

yellow face are apparent at a considerable distance.

Common Black-Hawk adult soaring. Often soars with tail only partially fanned; constantly flexes and adjusts wings.

IN FLIGHT. Adult Common Black-Hawk borders on unmistakable in flight. The wings are exceedingly broad throughout their length, disproportionately wider than any other North American raptor. Fully fanned, the tail is extremely short, abbreviated, barely extending behind the trailing edge of the wing. In fact, the bird often appears to be all wing — just a soaring wing with a head.

In overall shape, Common Black-Hawk is most like Black Vulture, but the wings are conspicuously wider, and Black Vulture lacks the single broad white tail band that is so conspicuous on adult Common Black-Hawk. (Adult Common Black-Hawk also has a small white comma at the base of the outer flight feathers, but as field marks go, it is a far cry from the prominent white patches that adorn the tips of Black Vulture's wings.) On a high-flying adult Common Black-Hawk, the narrow white terminal band is often difficult to see. Also, on high-flying adults, the broad white midtail band sometimes makes the dark, distal portion of the tail look disembodied — a dark band following the bird.

Zone-tailed Hawks are frequently encountered in Common Black-Hawk territory and, if perched, are candidates for confusion. On flying birds, the plumage may be similar, but the shape and behavior of the two species are very different. Zone-tailed Hawks have much narrower wings and much longer tails, and they soar with a marked dihedral (reminiscent of Turkey Vulture). Common Black-Hawk soars on flat or slightly drooped wings. Also, Zone-tailed shows a real contrast in the wings — silvery flight feathers trailing dark coverts. The flight feathers of Common Black-Hawk are paler than the coverts, but less silvery than the flight feathers seen on Turkey Vulture and further masked by the broad black trailing edge of Common Black-Hawk's wing.

Juvenile Common Black-Hawk is also striking, but very unlike the adult. With its cold charcoal brown upperparts, boldly patterned face, and heavily streaked breast and belly, it resembles (in plumage, not in shape!) a large, tawny juvenile Gray Hawk. In flight, juvenile Common Black-Hawks show pale wing panels on the primaries that are visible above and below (reminiscent of many juvenile Golden Eagles). The tail, while longer than the tail of adults, is still conspicuously short and shows numerous wavy dark-and-light bands ending with a wider terminal band.

Plumage similarities notwithstanding, juvenile Gray Hawk and juvenile Common Black-Hawk are not likely to be confused. Despite the longer tail, the overall shape of juvenile Common Black-Hawk is similar to the adult — which is to say, wholly unlike the smaller, slimmer-winged, longer-tailed, accipiter-like juvenile Gray Hawk.

Common Black-Hawk soars on flat wings, with wingtips sometimes curled up.

In a glide, the wings are slightly down-curved throughout their length. In a hard glide, the primaries are pulled way in and the outer wing (hand) angles sharply back. This, combined with the then-bulging S-shaped secondaries, confers a gliding shape very reminiscent of Black Vulture.

To maneuver while soaring, the Common Black-Hawk bends the wings at the wrist. Like eagles, they sometimes maneuver by flexing one wing at a time. In a full soar, the wings are thrust slightly forward. Now fully unfurled, the bulging secondaries give a graceful, arching curve to the trailing edge of the wing. It is in this posture that the birds seem all wing.

The flapping flight is lumbering, effortful; wingbeats are slow, strong, and measured, very reminiscent of Bald Eagle's flap, and not at all like the quick, choppy, hurried flap of Black Vulture. Like Bald Eagle, wingbeats rise and fall equally above and below the body. After prolonged flapping, glides are lengthy at both high and low altitudes. When it's gliding, the laws of physics seem suspended or at least slowed. Birds lose altitude less quickly than size, proportion, and logic would seem to dictate.

South of the border, through much of Mexico, Central America, and northern South America, the Common Black-Hawk needs to be separated from the closely related and very similar Great Black-Hawk (*Buteogallus urubitinga*). There are plumage differences, such as an extra white tail band near the body and white uppertail coverts on Great Black-Hawk, but the most obvious difference under field conditions is a noticeably longer tail, which is typically less spread in a soar than on Common Black-Hawk. Great Black-Hawk also has longer wings that may be just as wide yet appear narrower than those of Common Black-Hawk. While unconfirmed in North America, this species has been reported in the Rio Grande Valley and breeds approximately 150 miles south of the United States border.

1. An adult Common Black-Hawk soaring and banking. Note the single broad white tail band and the pale comma at the base of the outer two flight feathers. Most of all just look at the shape! Arizona, April NH

2. An adult Common Black-Hawk in a full soar. Short, very broad wings fairly envelop the tail, making the bird look like a blocky black barn door. Note the subtle touches of brown in the body. These touches can also be seen in the secondaries from above and below. Arizona, March NH

3. An adult Common Black-Hawk in a full soar (same bird as #2). Backlighting infuses pale flight feathers with a touch of brown, but the whiter commas near the tip of the wing remain prominent. NH

4. A juvenile Common Black-Hawk. Not exactly lanky but overall longer tailed and less compact than adult. Still, the broad-beamed, blocky proportions of Common Black-Hawk are obvious. Heavily streaked juveniles show pale wing panels (above and below) and multiple wavy bands on the tail. Mexico, Autumn WC

5. The high backlit silhouette of a gliding adult Common Black-Hawk. There just aren't a whole lot of birds that you are going to see in North America shaped like Common Black-Hawk, but if you are reminded of Black Vulture, there is good reason. Veracruz, October CS

6. An adult Common Black-Hawk in a shallow glide. Note how on the trailing edge of the wing, the wide and bulging secondaries almost seem to droop. Note, too, how the wing is flexing and showing a bend at the wrist. Sonora, Mexico, April LD

Zone-tailed Hawk

From the arid canyon lands of southeastern California (where it is rare) to the riparian cottonwood greenbelts and sky islands of Arizona (where it is reasonably common), to the prairie hills and cattle-country grasslands of western Texas (where it is uncommon and local), and on to the oases of the Rio Grande Valley in southern Texas (where it is again rare), the Zone-tailed Hawk enjoys a far wider range than the other borderland birds included in this section. With a North American population estimated to be approximately 150 pairs, Zone-tailed is numerically less common than Common Black-Hawk but, owing to its penchant for aerial hunting, more likely to be seen where the two species overlap.

South of the U. S. border, Zone-tailed Hawk is common throughout Mexico, Central America, and much of South America and is found in a variety of habitats. Like many buteos, Zone-taileds are active, aggressive predators of birds, lizards, and small mammals, many of which they secure at the end of sudden, almost recklessly fast stoops.

The Zone-tailed is a large buteo, overall smaller than Turkey Vulture, less bulky than Common Black-Hawk, about the size and wingspan of Red-tailed Hawk. However, Zone-tailed appears more narrow winged and longer tailed than Red-tailed, and in shape and demeanor it is more like Rough-legged Hawk.

The species that Zone-tailed Hawk most closely resembles is not a buteo, not even (by taxonomic standards) a raptor. Zone-taileds look and behave almost exactly like Turkey Vultures, and many authorities consider them vulture mimics. The supposition is that by utilizing the guise or behavior of vultures, Zone-taileds gain a strategic advantage over prey.

It is also argued that similarities in shape and manner of flight may be just parallel aerodynamic adaptations. Both species course low over the ground and along ridges searching for food (as do harriers, Ferruginous Hawks, and Golden Eagles).

Mimic or not, it cannot be denied that Zone-tailed Hawks look and act like Turkey Vultures. It may or may not confuse prey. It certainly confuses hawk watchers.

The Zone-tailed Hawk commonly hunts in a manner that combines the ridge-cruising hunting style of a Golden Eagle with the quartering flight of a Northern Harrier. This patrolling flight is usually higher than the flight of a harrier.

Zone-taileds also mimic the soaring flight of vultures (in fact, they join soaring and gliding vultures); however, once they sight prey, all resemblance to Turkey Vulture disappears. Stoops are dramatic, almost meteoric. A bird that feeds primarily upon carrion (like Turkey Vulture) has little need for speed. A bird of prey that is a bird-catching specialist does.

The Zone-tailed Hawk is easily the most migratory of the southwestern specialty buteos. While Gray Hawk and Common Black-Hawk withdraw south in winter, few are seen at hawk watches. Zone-tailed Hawks, on the other hand, are regularly seen in migration. They arrive in the United States in mid-March and head south by early October. They are seen migrating not only over the Rio Grande, at Big Bend, and in the Lower Valley each year, but are also regularly at watch sites such as the Sandia Mountains in New Mexico in spring and the Manzano Mountains in New Mexico in fall.

At Veracruz, Mexico, Zone-taileds are regular migrants. Up to 250 are recorded annually among the stream of hawks and vultures heading south in September and October.

Zone-taileds wander, too, and have been recorded in Nevada, Utah, Louisiana . . . even the Florida Keys and Nova Scotia! A few also winter in the United States (in south Texas and southern California).

If this species can appear as far outside its normal range as Nova Scotia, it can appear almost anywhere — in fact, it probably has! The problem is, unless you are actively looking for it, the bird will probably be dismissed as "just another Turkey Vulture."

IDENTIFICATION. Stated plainly, the Zone-tailed Hawk is a buteo that looks like a Turkey Vulture. Adult and juvenile Zone-tailed Hawks are dark, appearing nearly coal black under most field conditions, with pale, contrasting trailing edges to the wings. Juveniles have spare white spotting (primarily on the breast) otherwise, and in the hawk-watching arena, they are like the adult. On cloudy days, birds appear somewhat lighter, dark slate gray.

Juveniles also show a typical buteo outer wing panel, both from above and especially from below. This slightly lighter wing panel, faint translucent windows on the inner primaries, is visible at quite a distance (and something Turkey Vultures never show).

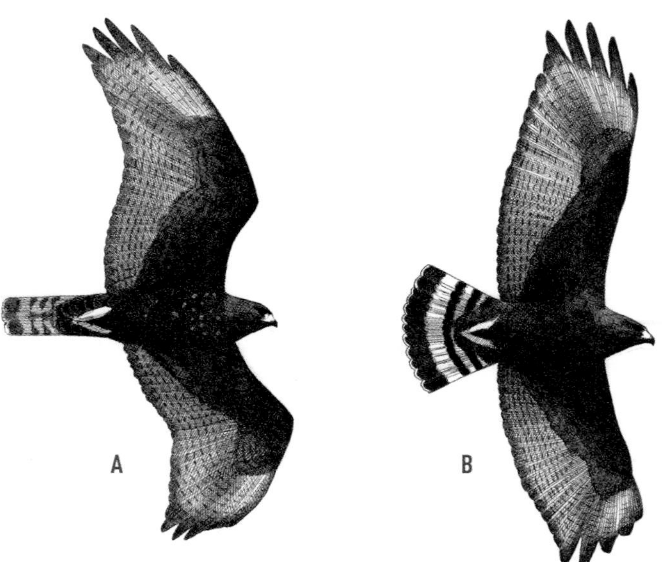

Zone-tailed Hawk, underside. Juvenile gliding **(A)**. Adult soaring **(B)**. Identification must focus on the overwhelming similarity of this species to Turkey Vulture, as these species are often seen together. Subtle differences in shape are useful indicators, but positive identification should be based on details (see Figure on p. 261). Juvenile basically identical to adult but with finely banded tail and usually some white spots on body. Other dark buteos can appear similar, particularly Rough-legged, which, however, has much whiter flight feathers. Red-tailed has broader wings; juvenile White-tailed has more pointed wings.

Legs are bright yellow, but this is rarely a useful field mark on flying birds (insofar as the legs of vultures are often caked with excrement and also appear pale). The cere, too, is yellow; the facial skin is gray and, while apparent, is not so prominent as the yellow on the face of Common Black-Hawk. When Zone-tailed is perched, its long wings reach the tip of the tail; on Common Black-Hawk, the shorter wings do *not* reach the tip of the tail (despite everything that has been said about Common Black-Hawk's short tail).

Zone-tailed Hawks show some sexual dimorphism. From below, adult males have two whitish tail bands (one wide, one narrow) and adult females three (one wide, two narrow), but you need to be very close to see this. In flight, observers commonly see one band on the male and perhaps two on the female. The others are hidden by the undertail coverts and muted by distance. Mainly, you get the *feel* of more white in the tail of the female. The whitish band is as wide as the band on the tail of Common Black-Hawk but not so starkly white. Seen from above, the bands look gray.

The tail of juveniles is dirty white and narrowly banded. At a distance, the tail appears dull and silver gray, as does the trailing edge of the wing.

The pale tail band(s) of Zone-tailed Hawk is not necessarily its most obvious or most useful field mark. Prominent only when birds are soaring close overhead, the bands are often hidden in the closed tail, giving the birds an overall pale-tailed effect (not unlike Turkey Vulture). Usually, one or more other distinguishing traits (to be discussed) will become apparent before the tail bands are discernible.

The female Zone-tailed is more broadly proportioned than the male. When mated pairs soar together, you can pick the slimmer male out by structure alone (no need to count tail bands). Males are rangier and more harrierlike; females are wider winged, proportioned more like Turkey Vultures.

A final plumage point: While Zone-tailed's pale flight feathers contrast with much darker coverts from below, these pale feathers lack the bright silver "mirrors flashing in sunlight" quality seen on Turkey Vulture. This muted contrast, caused by a patterned barring on Zone-tailed's pale flight feathers, is easily noted when Zone-taileds soar with vultures, but less apparent when birds are seen alone.

The barring itself may be a useful field mark. Present in Zone-tailed, absent in Turkey Vulture, this fine detail (even noted at long distances) is visible through the excellence of today's optics. But even the finest optical gear has its limits. At great distances and in poor light, a Zone-tailed's wings may simply appear all black.

IN FLIGHT. Zone-tailed Hawk appears large yet slim. The long, narrow, planklike wings and long narrow tail give it a rangy look — recalling more the straight edge of a Rough-legged Hawk or Northern Harrier than the chunky, curvaceous Red-tailed. The flight of Zone-tailed also recalls Northern Harrier. The movements are loose, leisurely, and languid; in high winds, they are tippy and jerky.

When soaring, the Zone-tailed holds its wings perpendicular to its body or thrust forward very slightly. The leading edge is almost straight-cut; the trailing edge curves ever so slightly (more so along the secondaries), but not so much that the wings lose their clean-cut, straight-edged appearance. By contrast, a soaring Turkey Vulture shows a slight bend or bulge at the wrist along the leading edge of the wing.

Zone-tailed Hawk, upperside. Adult soaring. Uniformly dark blackish with gray tail bands; white tail band is exposed when tail is more broadly fanned. Juvenile similar except with many narrow tail bands and never shows white in tail.

The Zone-tailed's tail is squared at the tip. The tail of Turkey Vulture is more rounded or lobed. Like Turkey Vulture (and unlike most buteos), Zone-taileds usually soar with their tail closed. Zone-tailed also soars with a pronounced dihedral that is very reminiscent of (but not identical to) Turkey Vulture, and this V-shaped configuration is apparent when birds are soaring or gliding. Viewed critically, the dihedral of Zone-tailed Hawk shows a modified dihedral. The arm is raised but the hand is flattened, somewhat like Rough-legged Hawk — but the arm is raised with more gradual curve than abrupt jut, and the dihedral is always more acute on Zone-tailed.

Well, almost always. During migration, perhaps because mimicry is not as important as movement, the dihedral on Zone-tailed Hawk is often diminished. The bird still recalls Turkey Vulture, just not so much.

As should be obvious by now, the challenge presented by Zone-tailed Hawk doesn't lie in separating this species from other southwestern buteos but from Turkey Vulture. At close range, structural and plumage differences easily distinguish the two species (the head of Turkey Vulture is small, red, and unfeathered, while the head of Zone-tailed Hawk is feathered, buteo-broad, and black; the tail of Turkey Vulture is wide and unmarked, while the tail of Zone-tailed is bisected by one or two distinctive white bands).

But distance and a peculiar alchemy relating to traits and relative proportions result in a structural homogenization. For example, while the head of Turkey Vulture is small and naked, the neck is large and thickly feathered. Distance transforms "naked head" into "pale bill," and "wide neck" assumes the properties of "large head," with the result that what projects ahead of the wing of both birds blurs into structural similarity.

Another example: The tail of Turkey Vulture is unmarked, while the tail of Zone-tailed Hawk is bisected by a broad whitish band. But Turkey Vulture shows pale feet and legs that fall precisely where the white band of Zone-tailed Hawk lies, with the predictable result that the banded tail of Zone-tailed loses its distinctiveness.

The tail of Zone-tailed Hawk is narrower than that of Turkey Vulture, as are the wings. But the relative proportions are similar, with the result that, at a distance, the overall silhouettes of the two birds are nearly identical.

Distinguishing Zone-tailed Hawk and Turkey Vulture is challenging but not impossible.

The transmutations of distance notwithstanding, the head of Zone-tailed

Zone-tailed Hawk (top) and Turkey Vulture (bottom) soaring and going away. Zone-tailed shows more arched inner wing and sharply upswept hand.

Adult Zone-tailed Hawk (left) with Turkey Vulture (right). Strikingly similar, but Zone-tailed is smaller and more buteolike, with more distinct S-curve on trailing edge of wing and less-rounded tail. Details are the most reliable field marks: Zone-tailed shows white tail band, dark trailing edge on wing, yellow feet, more rounded feathered head, and yellow cere.

Hawk always appears broader, darker, and overall more prominent than that of Turkey Vulture. More than any other characteristic, head shape serves to distinguish these two very similar birds in flight. Another clue that works well is the silvery gray bill and yellow cere of Zone-tailed. In bright sunlight, set against the dark head, it fairly gleams and can be seen at great distances.

Also, Zone-taileds frequently fly with their feet partially dangling. Black Vultures exhibit this behavior; Turkey Vultures almost never do.

A Zone-tailed Hawk has a heavy flap, labored and deep, very Turkey-Vulture-like, even though it is a much smaller, lighter bird. But keep watching and you might notice that the Zone-tailed's actions are quicker and more agile than those of a Turkey Vulture. Being smaller, Zone-taileds turn tighter and move with just a little more alacrity than their vulturine confederates. And while usually (but not always) tipsy like a Turkey Vulture as they wobble about the sky even in light winds, this rocking seems somehow contrived or controlled, the actions of a more aerodynamically stable bird trying to disguise this fact. During migration, Zone-taileds gliding high overhead hold rock-steady.

The flap of Zone-tailed is slightly less labored and not quite so deep as the flap of Turkey Vulture, but the cadence of the wingbeat is similar. In a glide, a Zone-tailed's wings are more pointed at the tip than are a Turkey Vulture's, without the slotted primaries showing prominently.

Study a Zone-tailed Hawk long enough and it starts acting very hawklike — gliding a lot, holding into the wind, then tacking back and forth, riding the wind in a way that brings to mind a low, hunting Ferruginous Hawk or a Northern Harrier. Yes, the flight of a Zone-tailed mimics that of a Turkey Vulture. But vultures are merely searching; Zone-taileds are hunting.

There are a couple of blatant behavioral giveaways. For example, a "Turkey Vulture" that suddenly folds up into a rocketing stoop probably is not. While vultures circle or glide down toward carrion, Zone-taileds first must make it carrion.

Vultures are generally ignored by other birds. Zone-taileds, which are consummate bird catchers, are not. An apoplectic Western Kingbird, tearing the feathers out of the back of a circling "Turkey Vulture," is urging you to take a second look.

Other guides have made the differentiation of Zone-tailed Hawk and Common Black-Hawk a concern. While similar in plumage, these two species differ dramatically in terms of behavior, shape, and GISS.

Behaviorally, Common Black-Hawk is a perch hunter; Zone-tailed hunts from the sky. In flight, Common Black-Hawk is large and lumbering, a flying blackboard. Zone-tailed is rangy and nimble, an aerialist in vulture clothing.

Other dark buteos may be confused with Zone-tailed Hawk. Swainson's Hawk has a similar dihedral but is not *usually* so tippy in flight and (of course) has a dark, not silvery, trailing edge to the wing. The very uncommon dark-morph Broad-winged Hawk is much smaller and gently droops its wings, never raising them in a dihedral. Dark-morph juvenile Red-tailed Hawk, with its finely banded tail, poses a potential identification problem, but the broader, more bulging or curvaceous wing shape distinguishes it from the straight-cut, clean-limbed Zone-tailed. Also, the wings of Red-tailed are not held in so pronounced a dihedral.

Just remember, Red-taileds (and Swainson's and Broad-wingeds) will never strike you as Turkey Vultures.

A dark-morph Rough-legged Hawk would be a more challenging separation from Zone-tailed, although for this to be a real-world concern, one of them would have to be well out of its respective geographic and temporal range. The overall shape is similar, and both birds fly with a dihedral. The pale, lightly barred trailing edge of an adult Rough-legged's wings have the tarnished-silver quality of a Zone-tailed's, and the barred tail pattern of an adult dark-morph Rough-legged resembles that of a juvenile Zone-tailed.

Mainly we come down to GISS. While Rough-legged Hawks may rock modestly in flight, particularly in a brisk wind, they are nowhere near as unsteady or tippy as Zone-tailed Hawks (or Turkey Vultures!).

In the final analysis, the challenge inherent in the identification of Zone-tailed Hawk comes down to telling apart the Turkey Vulture and the Turkey Vulture mimic. If you find it difficult, don't despair. Fledgling Zone-tailed Hawks have been known to open their mouths and scream loudly to every passing Turkey Vulture, begging to be fed.

See? It's not just us.

1. An adult female Zone-tailed Hawk showing three grayish tail bands (in many cases, the third inner band is obscured). This bird is soaring, but it is also actively hunting, as the intensity of its gaze attests. Texas, September NH

2. An adult female Zone-tailed Hawk soaring. Note the long wings, pronounced dihedral, pale trailing edge to the wing, but showing a banded tail — a feature never seen on Turkey Vulture. Arizona, July NH

3. An adult male Zone-tailed Hawk in a glide, carrying prey. Males show only two bands in the tail. In this configuration, gliding and alternately flapping, the bird appears quite buteolike and far less like a Turkey Vulture. Arizona, May NH

4. An adult male Zone-tailed Hawk in a steep glide, with the wings held in a Turkey Vulture–worthy dihedral and angle. The bright yellow cere is evident at surprising distances. Arizona, June NH

5. A soaring juvenile Zone-tailed Hawk, head-on, showing multiple narrow dark bands on the tail, more patterned flight feathers, and tiny white flecking in the dark body and underwing linings. Note the very prominent dihedral. Arizona, July NH

6. The same bird as in #5, in full profile. Notice the flecking on the body and the finely patterned flight feathers. This is one stunning hawk at close range, where it will appear very un-Turkey-Vulture-like. NH

7. The dorsal view of an adult male Zone-tailed Hawk. Note the very pronounced Turkey-Vulture-like dihedral, but also the un-Turkey-Vulture-like gray plumage (not to mention the bands in the tail). Zone-taileds often, but not always, rock and tip in flight much like Turkey Vultures. New Mexico, April NH

8. Things look different at a distance than they do up close. A soaring adult Zone-tailed Hawk. Zone-taileds soar with tail closed (as Turkey Vultures commonly do) and can appear as small headed as Turkey Vultures as well. Veracruz, October CS

9. An adult Zone-tailed Hawk in glide. You can see the similarities between this bird and Turkey Vulture, and you have to look closely to note the differences: less contrast between the underwing coverts and the flight feathers, semblance of a band in the tail, pale cere and bill gleaming in the sunlight, and a head that is just too wide and prominent for a vulture (at least from this angle, if not in #8). Veracruz, October CS

White-tailed Hawk

The Texas Coast is large and flat. Only the heat waves interrupt the view to the horizon. In winter, it is not only featureless, but colorless, too—just browns and grays; grass and mesquite, with some prickly pear thrown in.

But for the birder, the Texas Coast hosts an assortment of special birds. One of the best is the White-tailed Hawk. As Florence Bailey so eloquently described in 1916 in the *Condor,* "In driving over the treeless prairie of southern Texas day after day, one of your keenest pleasures is to sight, across a long, level foreground, the shining breast of the stately White-tailed Hawk." Indeed, the White-tailed Hawk not only helps make the sameness of the coastal bend palatable, it makes it exciting.

The White-tailed Hawk is the most eastern of the southwestern buteos. There are old records from Arizona and New Mexico (likely birds from the Mexican, not the Texas, population) and irregular winter sightings in extreme southwest Louisiana. But in the United States the White-tailed is a Texas resident, common, albeit locally distributed, from Falcon Lake to Brownsville on the Mexican border, north to Houston and Galveston Bay.

It haunts semiarid to arid open chaparral, prairie, and cord grass flats along the length of the coastal bend. It is easily seen at places like Attwater Prairie Chicken National Wildlife Refuge, Aransas National Wildlife Refuge, and Laguna Atascosa National Wildlife Refuge. These birds appear on Padre Island in winter but do not breed there. Coastal in nature, White-tailed Hawk is rarely seen 75 miles inland.

The birds are easily found, yet they are not common; the population of White-tailed Hawks in Texas may not exceed 400 to 500 pairs. Officially designated a threatened species (largely because of its restricted range), the Texas population is nevertheless stable or may be growing.

South of the U. S. border, White-tailed Hawks are common on both coasts of Mexico wherever proper habitat is found. They are also common raptors of the vast Venezuelan *llanos* in the northern and central parts of the country. Of interest, the White-tailed Hawk lays claim to the widest latitudinal permanent distribution of any buteo (20° N to 44° S).

In the United States, the White-tailed is nonmigratory. This has obvious implications for the traveling hawk watcher, who can look for the White-tailed and its compatriot, the Harris's Hawk, in winter (unlike the other species in this chapter).

In winter, White-tailed Hawks have established a pattern of concentrating near coastal lagoons, then dispersing inland in the spring to scrub- and live-oak areas where they breed. Somewhat nomadic and opportunistic, the birds are attracted to prairie fires. Up to 60 individuals have been seen over such blazes, feeding upon prey flushed by the flames (and the birds have been known to follow tractors for much the same reason).

Prey is eclectic. Pocket gophers, rats, mice, and small rabbits predominate, but the birds commonly take snakes, lizards, and frogs, too. Carrion also figures in the diet, and they have been known to haunt chicken ranches, where discarded dead poultry is the attractant.

In some studies, up to a third of the White-tailed's diet has been shown to be

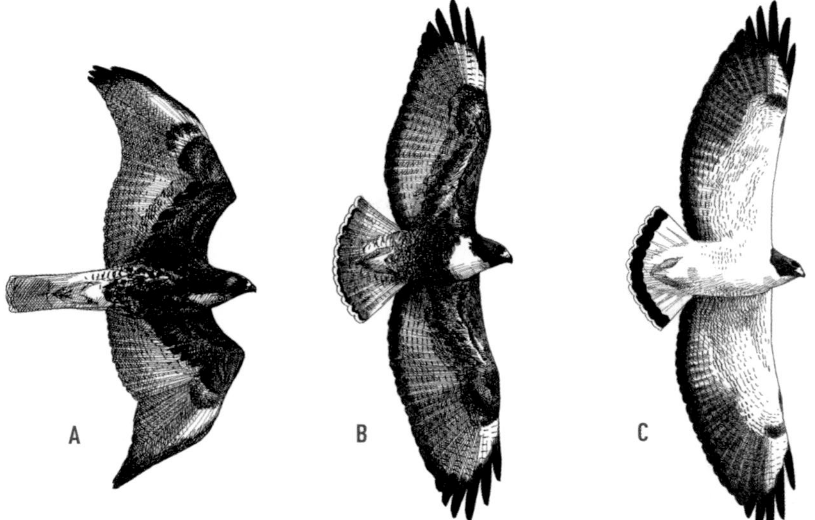

White-tailed Hawk, underside. Juvenile gliding **(A)**. Subadult in nearly full soar **(B)**. Adult soaring **(C)**. Elegantly shaped and patterned, this species shows the most striking age-related differences in shape of any North American hawk. All have long, tapered, pointed wings pinched in at base. Adults have wings broad on secondaries, tail quite short. Juveniles have wings narrower, more flowing, less tapered, and tail longer. Adults fly with more pronounced dihedral than juveniles. Subadults are variably intermediate in shape. Adults have striking clean white body and tail, dark gray inner primaries. Juveniles have brownish gray body and underwing coverts with variable creamy whitish patch on chest, mottled underwing coverts, dull gray flight feathers, and pale grayish tail. Subadult is similar to Red-tailed Hawk but with wings more pointed at tip and more pinched at base, and tail pale grayish.

White-tailed Hawk, upperside. Juvenile soaring **(A)**. Adult soaring **(B)**. Adult striking with white tail and rump, pale gray back, and pale rufous shoulders. Juvenile uniformly dark brownish gray with neat white uppertail coverts and pale grayish tail, unlike any other species. Note also small white cheek patch on juvenile.

small to medium-sized birds, a difficult specialty for a buteo, especially one as large as White-tailed Hawk.

IDENTIFICATION. The White-tailed Hawk is a large, active, and handsome buteo, the same size as a Swainson's Hawk — a bird with which it also shares some structural and plumage similarities. While White-tailed Hawks do hunt from perches, this active aerialist is far more likely to be seen overhead, hovering and kiting in search of prey. Whether they are seen perched or flying, few would argue that White-taileds do not rank among our most striking birds of prey.

For adults, plumage is an easy route to identification. Above, the head and back of an adult are pearl gray. The wings are dark, almost black, emblazoned with a striking chestnut sash that adorns the upperwing coverts. The white of the tail extends up the rump, forming a white wedge up the back — a raptorial dowitcher!

From below, the body is gleaming white, arguably the whitest of all North American hawks. Only the rusty-backed Ferruginous is so extensively white below. Very fine black barring is etched across flanks and belly, but these markings are difficult to see in the field. The tail, too, is brilliant white above and below, with a crisp, wide, black subterminal band (bordered by a narrow white terminal band). It suggests, somewhat, the tail pattern of a young Golden Eagle.

The underwing coverts are also white (but not so gleaming). The flight feathers (in fact the whole trailing edge) are dark gray to blackish, imparting a two-toned wing pattern also found on Swainson's and Short-tailed hawks. If viewed at close quarters, you might see that the darkness of the flight feathers is the product of tightly packed barring (but under most field conditions, the trailing edge just seems dark). Secondaries are paler than primaries — a pattern that is just the opposite of Osprey and Northern Harrier and adds to the adult White-tailed's uniqueness.

Juvenile White-tailed Hawk looks very different. Above, its blackish brown body and wing coverts contrast with slightly paler, grayer flight feathers. The head is patterned with white patches over the eye and on the cheek (a pattern that recalls a very dark Prairie Falcon's face). The tail is pale gray, welded to the body with a prominent white U-shaped patch (recalling Swainson's Hawk or young Gray Hawk).

Below, juveniles are mostly dark sooty brown, but variously mottled. All birds show a prominent white patch on the breast (a characteristic shared only by Harlan's Red-tailed Hawk). Many show a pale wing strut (somewhat like first-year Bald Eagle) sandwiched between the darker leading coverts and the dark trailing edge of the wing. In good light, the dark areas seem rust stained in places.

Unlike most other hawks, White-taileds have a distinct subadult plumage that is a blend of adult and juvenile traits. Above, subadult White-taileds are overall dark, much like juvenile birds — *except* for an all-dark head, russet shoulders, and dark-tipped, dusky gray tail. Confusion with Harlan's Red-tailed is possible (if you overlook the russet shoulders). Below, and except for the dark-tipped tail and overall darker patterning, subadult birds are much like juveniles — with their white chest and dark middle, they can look much like dark Red-tailed Hawks.

More bad news: Records of dark adults exist, making confusion with dark-morph Red-tailed, Swainson's, and Ferruginous hawks a possibility. But a dark morph adult still has a white tail.

White-tailed Hawk. Top to bottom: adult kiting, adult soaring, adult in level flight. This species kites frequently and expertly. Soars with dihedral on broad but pointed wings.

IN FLIGHT. The White-tailed Hawk is not only elegantly patterned, it is elegantly shaped. In fact, the adult White-tailed Hawk has one of the most unique shapes of any North American hawk.

Among North American buteos, the adult White-tailed has the longest wings and (except for Common Black-Hawk) the shortest-appearing tail. The wing is uniquely configured. The hand is very long and acutely tapered, but the inner primaries are shortened and the secondaries bulge — so much, in fact, that the spread tail of a soaring bird touches, even overlaps, the trailing edge of the wing. The net effect is idiosyncratic: *all wing, no tail.* In overall shape, the silhouette of an adult White-tailed Hawk recalls the shape of the Bateleur, *Terathopius ecaudatus,* a savanna- and thorn-scrub-loving raptor found widely across sub-Saharan Africa.

Immature White-taileds are a whole other story. In fact, in terms of shape, they seem like a different species. While long and tapered, the wings of juvenile and subadult White-tailed Hawks lack the bulging secondaries. The wings are overall narrower, more evenly and more uniformly tapered, more . . . Swainson's Hawk–like!

The tail of a nonadult White-tailed Hawk is also considerably longer. Combine this with the narrower wings, and the tail on young White-taileds appears twice as long as the stubby-tailed adult's. Once again, the overall shape is Swainson's Hawk–like and very unlike the Bateleur-harking shape of the adult White-tailed Hawk.

White-tailed Hawks soar and glide with a pronounced dihedral — more acute than the dihedral of a Swainson's Hawk but not quite as pronounced as a Northern Harrier. Adults, however, fly with a more pronounced dihedral than young do.

In a glide, White-taileds appear hunch shouldered with sharply swept-back wings. The tapering on the wingtips becomes acute. On gliding and kiting birds, wingtips curl back and look like sharpened pencil points snapped back at an angle. In a glide, the birds rock back and forth — the movement is not floppy like a Turkey Vulture's, but tippy like a hunting harrier's. When hunting low, through brush, they can be surprisingly agile and dexterous.

Energetic and active fliers, White-taileds typically hunt aloft. With the exception of Short-tailed Hawks, they kite and hover with more finesse than any other buteo. Kiting White-taileds hold their wings high over their back, suggesting springboard divers ready to plunge. The wings are motionless except for the tips, which make subtle adjustments as they micromanage the air.

Adult White-tailed Hawks, with their seemingly all-wing profile, striking plumage, and within the confines of their limited range, are virtually unmistakable. Juvenile and second-year White-taileds, however, have plumage, shape, and flight characteristics that make them and juvenile Swainson's Hawks easy candidates for confusion. Both have long tapered wings and long tails. Both have a dark trailing edge to the wing. Both have white patches at the base of the tail, and many Swainson's Hawks have bold facial patterns and some white on the breast (usually relegated more toward the throat).

The white rump and breast patch are more pronounced on juvenile White-tailed. Also on juvenile White-tailed, the tail, while finely vermiculated, appears gray and unbanded (except for a single dark tip on second-year birds); the tail of Swainson's Hawk is narrowly banded. Overall, young White-taileds are more patterned or mottled below but cleaner and less spangled above.

Juvenile White-taileds are similar to Swainson's Hawks not only in plumage but also in shape. However, while both birds have long tapered wings, the wings of juvenile and second-year White-taileds are slightly more curvaceous — tapered in the hand, wider in the arm, and often pinched in where the wing meets the body. The wing of Swainson's Hawk is more slender, more evenly tapered and clean lined, candlestick-like. The dihedral is also very similar, although White-taileds (particularly adults) usually fly with a dihedral that is more acute.

One distinct difference is the soaring profile. Swainson's Hawk soars with wings held perfectly perpendicular to the body; the leading edge of the wing is straightedge flat. White-taileds soar with wings thrust slightly forward — not as much as Red-shouldered Hawk, but still apparent. The leading edge bulges at the wrist and often the hand curves gently back. Swainson's Hawks curve their hands back only when gliding, not soaring.

In flight, White-tailed appears heavier than Swainson's. Wingbeats are slower, deeper — more like a Red-tailed Hawk than a Swainson's Hawk.

As suggested earlier, confusion between young White-tailed and dark-morph Red-tailed (particularly Harlan's) is a possibility. On the underparts of some subadult White-taileds, the white breast and white undertail coverts forge a belly-band effect. Dark and intermediate forms of Red-tailed Hawk have dark undertail coverts (not white), and on second-year White-taileds, look for the hint of chestnut on the shoulders. Shape is the most distinguishing difference. Red-taileds have shorter, broader, and much less tapered wings and fly with less dihedral.

These characteristics also distinguish light-morph Short-tailed Hawk from White-tailed, should either of these species appear out of their normal north-of-the-border range (and they have). From below, adult light-morph Short-tailed Hawk and adult White-tailed Hawk have a similar plumage pattern, except that Short-tailed has a multibanded tail, not a white tail tipped with a paintbrush-wide black swath.

While the shorter and stocky-winged Short-tailed Hawk is equally a master of kiting and stop-and-go aerial control, soaring and kiting Short-taileds fly on flat wings, not the dihedral of a White-tailed. Only the wingtips of Short-tailed are swept up.

1. A soaring adult White-tailed Hawk showing brilliant white underparts contrasting with the dark trailing edge of the broad wings. The wings taper acutely toward the tip. Texas, December NH

2. A soaring adult White-tailed Hawk showing the pronounced dihedral that is so characteristic of this coastal prairie species. Texas, December NH

3. An adult White-tailed Hawk in a shallow glide. While the wings on this species are exceedingly long, a fully fanned tail alters the shape — in fact, the tail seems to fuse with the wings. The tail on this floating bird is only partially fanned. Texas, September CS

4. The dorsal view of a subadult White-tailed Hawk, flapping in direct flight. It is showing a mix of age traits — the reddish shoulders, grayish wing, and black-tipped white tail hark to adult, while the heavily barred underparts are a juvenile trait. Also, two brown juvenile secondaries are visible on the far wing. Texas, December NH

5. A subadult White-tailed Hawk in a shallow glide. This age can superficially resemble Red-tailed Hawk, with a dark head, white chest, and dark belly band, but note the very long and pointed wings, dark gray flight feathers, and lack of dark patagium. Texas, December NH

6. A fairly typical juvenile White-tailed Hawk. The wing shape (in fact, the entire silhouette of the bird) differs subtly from that of adults. Juveniles appear rangier, more slender winged and longer tailed (overall more like Swainson's Hawk). In fact, if you saw this bird next to an adult White-tailed, you might think you were looking at different species. Texas, December NH

7. A dark juvenile White-tailed Hawk (on the very dark end of the spectrum) in a glide. This bird shows a small white spot on the breast but also, as is typical, a touch of white on the sides of the face (see #6 for the white eyebrow, which is also typical). Texas, December NH

8. Same bird as #3, angling away. This angle best illustrates and accentuates the acutely tapered wing shape of White-tailed Hawk. The wing shape is as much a signature field mark as the white tail. CS

Harris's Hawk

If the raptors in this chapter were to elect a representative from the borderland, Harris's Hawk would be the ideal candidate. Tipping-point considerations would include the bird's plumage, an earth brown body emboldened with black and white and red — a Navajo blanket in feathers. Compelling, too, is the encompassing scope of Harris's Hawk's range. More widespread than the other species in this section, Harris's Hawk is found along the Colorado River of California; in the deserts of Arizona, eastern New Mexico, and southern Texas; and all the way east and north to just above Corpus Christi.

A 1980s winter population estimate suggested that about 5,500 Harris's Hawks were found in the United States. That's the good news. The bad news is that the number from a decade earlier was 10,000 birds. While more numerous and more likely to be seen than Gray, Common Black, Zone-tailed, and White-tailed hawks, the population and range of "Bay-winged Hawks," as the bird was formerly known, have been steadily shrinking.

Outside the United States, Harris's Hawk is widely distributed and quite common throughout Mexico, Central America, and much of South America, south to Chile and northern Patagonia. The species is nonmigratory, but family groups can be nomadic or even irruptive in search of food. Harris's Hawks are intelligent and determined predators, agile and dashing hunters of mammals and birds. In the American Southwest, they prey heavily on desert cottontails but kill jackrabbits as well. Gambel's and Scaled quails are also important prey.

The most intriguing aspect of Harris's Hawk biology is its social nature. Rarely does an observer encounter a single individual. These raptors both nest and winter in social units of up to six or seven birds, usually an extended family group in which a previous year's young share parenting duties for the next generation. Harris's Hawks may be monogamous, polygamous, or polyandrous. They are well known for their cooperative foraging. Just like human hunters, there are "beaters" and "standers." One or several will enter thick brush, sometimes on foot, flushing prey in the direction of hovering or strategically perched kin. In one of their most celebrated (and photographed) maneuvers, Harris's Hawks will perch atop one another, forming raptorial totem poles, gaining height and a better view.

It is this social proclivity (as well as their hunting skill) that endears Harris's Hawks to falconers. Harris's Hawks appear to accept a handler as a partner in the pursuit, a partnership that leads to a social bond between hawk and handler. In keeping with the totem-pole tradition, Harris's handlers commonly wear hats, offering their birds the highest available hunting perch in perch-poor habitats.

Note: Because of Harris's popularity among falconers, escaped Harris's Hawks account for most of the out-of-range sightings of this species across the United States and Canada.

When chasing prey, a Harris's Hawk is dashing, bold, and determined. Its longer-than-buteo tail confers upon the bird something of the agility and GISS of an accipiter.

Though social, active, and (in places) common, Harris's can, at times, be difficult to find. The species spends hours perched and preening. Whether down in

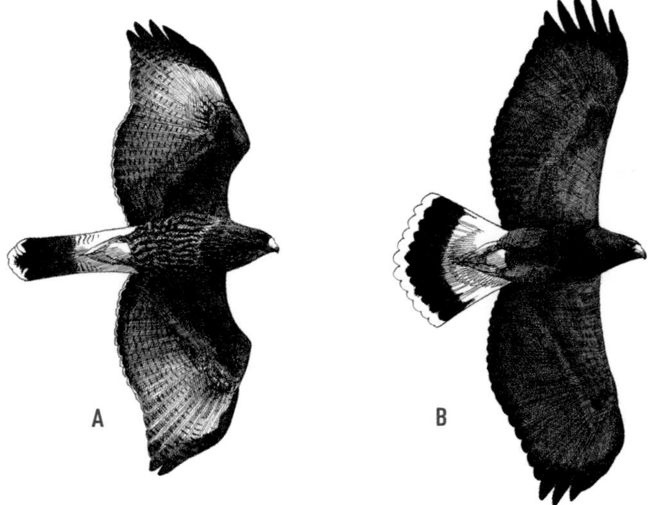

Harris's Hawk, underside. Juvenile gliding **(A)**. Adult soaring **(B)**. Long bodied and broad winged. The wings are rounded, very broad on the secondaries, and pinched at the base. Tail long and rounded. All ages show striking white patch at base of tail contrasting with dark body plumage.

Harris's Hawk, upperside. Adult gliding. Striking and distinctive plumage: dark brown with bright rufous shoulders, white rump, and black tail with white tip.

mesquite/creosote brush land, or high and still in the shadows of a streamside cottonwood, it's all the same: the birds are often out of sight. Happily and more often, the birds are prominently perched (often on utility poles) — easy to find, easier to identify.

IDENTIFICATION. Identifying the Harris's Hawk is straightforward. It is a large buteolike hawk, slightly smaller and leaner than the average Red-tailed — a decathlete among raptors, not a shot-putter. The wide wings, bulging inner primaries and secondaries, plus the long tail more easily recall Red-shouldered Hawk.

An adult Harris's Hawk is tricolored — blackish, reddish, white. The body is sooty brown. The upperwing coverts and wing linings (and legs) are rich russet. The distinctive and boldly patterned tail, black with a brilliant white terminal band and a wide white base, is visible above and below. Only the Snail Kite shows anything approaching this tail pattern, but many dissimilar traits (not to mention the Gulf of Mexico) separate these two species.

The plumage of the juvenile Harris's Hawk is similar to an adult's — overall

Harris's Hawk adult in level flight, with broad floppy wings, long body, and stout tail.

Harris's Hawk soaring head-on, wings nearly flat but showing distinctive bulge on trailing edge of inner primaries.

dark, but heavily streaked on the chest and belly (while adults are solid brown). Unlike adults, juveniles show pale panels in the outer wing (like the windows on a Red-shouldered Hawk, but wider and more extensive). The tail is finely barred, not black, yet still white at the base. In terms of overall proportions, adults and juveniles differ little.

IN FLIGHT. Distance distills the Harris's Hawk's plumage to bare essence. At a distance, the white tail tip is not easily seen (although the white rump remains very evident). The chestnut coloration disappears, too. Birds appear all dark.

Fortunately, the Harris's has a distinctive shape. It has broad wings that are very rounded at the tip and pinched at the base. Bulging inner primaries and secondaries add bulk, making the wings conspicuously paddle-shaped. Birds soar with wings thrust slightly forward, evoking the forward-reaching configuration seen on a soaring Red-shouldered.

Harris's has a long tail for a buteo-type hawk and at close range can appear very rangy and accipiter-like (especially juveniles). The tail is slightly wedge-shaped, recalling a raven. At great heights, however, in a full soar, tail length diminishes — the white tip disappears and the bulging trailing edge of the wing masks the extra-long buteo tail. All sense of ranginess disappears and Harris's becomes a compact, short-tailed, paddle-winged buteolike bird that can look deceptively like a dark soaring Red-tailed Hawk (although under most conditions, the white basal portion of the tail is clearly evident).

At great heights, the bird appears headless, all wing and tail. The white basal portion of the tail disappears into the sky, making the dark outer portion of the tail look like a disembodied section following the bird.

For a bird that spends much of the day sitting around, the Harris's soars brilliantly on wings that are held flat or show a very slight dihedral — pretty much what a Red-tailed does. Unlike Red-taileds (and except when soaring), Harris's Hawks fly with drooped or cupped wings — with wings actually bowed (raised at the body and bending down from the wrist to the tip). The bowed wings, apparent when the bird is both gliding and flapping, recall Red-shouldered Hawk, but the bow is even more exaggerated.

The flap is shallow and graceful for a large hawk, akin to a Red-tailed's but a bit quicker. Individuals sometimes execute a hurried series of flaps when entering a thermal. Glides are buoyant, slow, and often protracted. When gliding, birds also may execute a wingtip droop similar to Turkey Vulture's.

Active hunting flight is more energetic than that exhibited by most buteos, more accipiter-like (Cooper's Hawk–like, to be precise). Pursuit through brush seems reckless (and pursuit of prey may continue on foot).

Harris's will hunt from a perch. Stoops from perches are fast, sometimes spiraling, headfirst, and, at times, vertical. But a hunting Harris's also hunts like a harrier as it beats the bushes, actively searching for (or blindly hoping to flush) prey. It also hunts aloft, at times kiting very well with wings thrust slightly more forward than the wings of a kiting Red-tailed Hawk. At times it seems to do stalls while gliding, kiting briefly before resuming the glide, but hovering is rare. In a high wind, a Harris's Hawk may quarter back and forth above the trees like a hunting Ferruginous Hawk.

Identification of Harris's Hawk is not usually difficult. The shape (and overall dark color) is somewhat akin to juvenile dark-morph Red-tailed Hawk, but a critical look will certainly disclose the anomalous, un-Red-tailed-like, paddle-shaped wing. Also, the white-dark-white tail pattern is determining.

A final, subtle point, but one that is surprisingly apparent to observers who have the basic buteo search image locked in their minds: It might be a matter of head projection, it might have something to do with the broadly dominating wings, or it might be something about the tail (or a combination of all these elements), but soaring Harris's Hawks give the distinct impression that the wings are set too far forward on the bird. Not awkwardly so, just curiously so — and it does get your attention.

1. An adult Harris's Hawk in a glide. When the bird is gliding, wings are slightly down-bowed, although this is only hinted at in this photo. Harris's is, for all intents and purposes, a buteo, and a classy one at that. Arizona, January NH

2. An adult Harris's Hawk banking. Note the tricolored plumage pattern — rufous, black, and white. At a distance, and except for the tail, the bird is all dark. The distinctive tail pattern shows a dark tail with a broad white base and white terminal band. Arizona, January NH

3. A juvenile Harris's (with streaked underparts and banded tail) soaring. Despite juvenile characteristics, the overall plumage pattern is like that of adults. Silhouette is classic, showing paddle-shaped wings and long tail fanned. Texas, February NH

4. A soaring juvenile Harris's Hawk showing classic broad wings and pale panels toward the wingtips (not visible from above) that are absent on adults. Texas, March NH

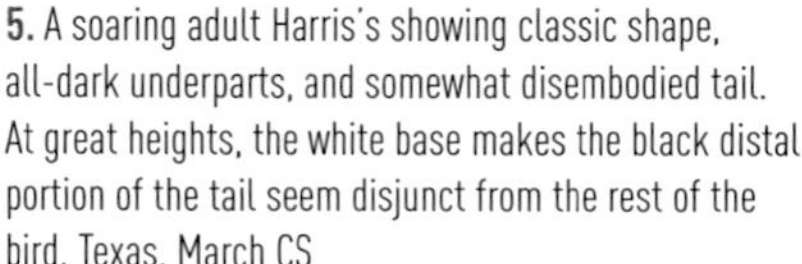

5. A soaring adult Harris's showing classic shape, all-dark underparts, and somewhat disembodied tail. At great heights, the white base makes the black distal portion of the tail seem disjunct from the rest of the bird. Texas, March CS

6. The dorsal view of a banking adult Harris's Hawk, from behind. The striking pattern of this beautiful hawk is evident at great distances (also see p. 242). Texas, March CS

7. Two adult Harris's Hawks soaring. Note the mostly flat arms and slightly uplifted hands of the upper bird and uniformly dark wings and body of the lower bird. Note, too, the broad, fully spread angular tail. Texas, March CS

PUTTING IT TOGETHER: Telling Southwestern Buteos Apart

The borderland buteos are a diverse group. For the most part, there is little problem distinguishing one from another. Only the Common Black-Hawk and Zone-tailed Hawk resemble each other in plumage, but the shape and GISS put them on opposite ends of the spectrum. More of a problem is separating the southwestern specialties from other, more widespread raptors.

Gray Hawk can resemble Broad-winged Hawk. Juvenile White-tailed may be confused with Swainson's Hawk. A distant or poorly seen Harris's Hawk can be mistaken for a dark Red-tailed. The greatest identification challenge focuses on distinguishing the Turkey Vulture mimic, the Zone-tailed Hawk, from the real thing. All these possibilities for confusion have been discussed in the individual species treatments.

The calendar can and should be an aid in making your ID. Gray, Common Black, and Zone-tailed hawks are migratory and, except for Gray Hawks in the Rio Grande Valley, withdraw from the United States in winter. White-tailed and Harris's hawks are sedentary and can be located year-round, but the juvenile White-tailed lookalike, Swainson's Hawk, spends the winter in Argentina.

Knowing what to expect and when is a great asset in the hawk-watching arena. The most expeditious way to eliminate possibilities en route to a correct identification is to not consider them in the first place.

However useful the marshaling of variables might be, observers should always be mindful, too, that hawks can and do sometimes appear outside both their geographic and temporal ranges (for example, a Zone-tailed Hawk in Nova Scotia on September 24; a Common Black-Hawk recorded in Minnesota). Harris's Hawks (mostly escapees) can be found all over the place.

The best thing, the important thing, is to seek out and become familiar with these wonderful and geographically restricted birds within their home range. This way, if and when they appear elsewhere, you will be ready.

Florida Specialties

The Aerialist and the Specialist

SPECIES
Short-tailed Hawk, *Buteo brachyurus*
Snail Kite, *Rostrhamus sociabilis*

There are many reasons to visit Florida — among them, Crested Caracaras, Burrowing Owls, Swallow-tailed Kites, White-tailed Kites, and the vast clouds of vultures that add character to the cumulus-filled sky. But, for the raptor enthusiast, there are two additional and very special reasons to visit the state. Florida is the only place in the United States where one can find and savor the Snail Kite and almost the only place one can study the Short-tailed Hawk.

Both are tropical species at the northern limits of their range. Both engage in population shifts that disperse and concentrate them over the course of the year. But two more dissimilar raptors could hardly be found to share the same chapter.

Short-tailed Hawk is a dive bomber of a raptor and in the air a master of stop-and-go control. It often hunts from heights that make it difficult to detect—and this, of course, is precisely why. If you want to locate Short-tailed Hawk, look up. Way up. And be grateful for the backdrop of cumulus clouds that enrich Florida skies.

Snail Kite is a hunting hound of a bird, searching low and slow over the "river of grass." "Stoop" seems not to exist even as a concept in the mind of Snail Kite and it is horizon, not the heights, that defines Snail Kite's workplace. But few birds maneuver so deftly or seem more at home in the Everglades environment as this hawk, whose range and diet are fixed by a freshwater mollusk. Find apple snails and you'll find kites. The image opposite does nothing but underscore this link and Snail Kite's finesse.

Neither bird is particularly difficult to identify, but, owing to their very small populations, locating Snail Kite and Short-tailed Hawk may present a challenge, albeit no great hardship for those fleeing the snow and cold that cover much of North America during the winter months. If those seeking Snail Kites and Short-

LEFT: An adult female Snail Kite over the Florida Everglades. The apple snail is the best supportive field mark. KK

tailed Hawks are forced to spend a few more days in the sunny bird-rich climes of southern Florida, the search will be no less gratifying than the prizes.

Florida hawks: soaring juveniles of all species, backlit. These three species are entirely unrelated. Short-tailed Hawk is a typical buteo. Snail Kite and Crested Caracara are both unique and distinctive. Crested Caracara **(A)**: long straight wings, long tail and long head, rowing wingbeats; all blackish with white head, tail, and wingtips; unmistakable. (*Note: this species is treated in Chapter 8.*) Snail Kite **(B)**: broad floppy wings held cupped and pushed forward when soaring, extremely long primaries flex in flight, tail short and square with obvious white base. Short-tailed Hawk **(C)**: long primaries and distinct bulge on secondaries with wings slightly pinched at body separate this species from Broad-winged and Red-shouldered hawks. Flight feathers rather dark, especially on secondaries, but some translucence shows on inner primaries.

Short-tailed Hawk

The Short-tailed Hawk is a tropical buteo reaching its northern limit in south and central Florida. It is widespread in Mexico, Central America, and South America — found south to Chile and northern Argentina. In Mexico, it is common north to Puerto Vallarta and, on the east coast, through Veracruz and Tamaulipas. Vagrants drawn from this Mexican population occasionally wander to southern Texas and Arizona.

Since the turn of the millennium, one or more pairs of Short-tailed Hawks have taken up residence in the Huachuca Mountains of Arizona, and juveniles have been recorded. This pioneering effort, as well as recent sight records from the Rio Grande Valley in Texas, may well indicate a range expansion.

For now, and for the opportunity to study multiple individuals of both color morphs, you'd do well to concentrate your search efforts in Florida. Short-tailed Hawks are found from the Keys north to the base of the panhandle, where they are uncommon and local nesters in mixed woodlands and savanna areas. Over most of its New World range, Short-tailed Hawk is nonmigratory, but in Florida, from mid-October through February, the United States population withdraws to southern Florida (south of Lake Okeechobee).

Short-tailed Hawks are sometimes seen in migration; 8 were recorded in Everglades National Park on November 1, and 11 were mixed in with Broad-wingeds at Key West on October 31. There is also a Dry Tortugas sighting (although, curiously, this widespread tropical species is not found in Cuba).

The entire Florida population is small, thought to be less than 200 pairs. When birds disperse across the state for the nesting season, finding Short-taileds can be a needle-in-the-haystack proposition.

It is in winter, in the Everglades and the Florida Keys, that Short-taileds are most easily found. Up to ten have been recorded on the Everglades Christmas Bird Count, and it is not hard for the focused hawk watcher to see four or five in a day.

Short-taileds are almost always seen in flight. When perched, they are secretive, usually deep in low to medium vegetation or tucked well within the leafy confines of a large tree — very unlike most other raptors, which choose conspicuous perches as platforms from which to locate prey.

IDENTIFICATION. The Short-tailed Hawk is an active hawk — a fast and feisty aerialist. Perhaps more than any other hawk (and certainly more than any other buteo), the Short-tailed can be identified by its behavior.

Short-tailed is almost exclusively a hunter of birds, a specialization that may be unique among the world's buteos. Unlike accipiters, Short-tailed hunts birds from above. It may spend hours aloft, often at heights above the limit of the unaided human eye, holding into the wind, balancing the forces of gravity and lift with the tilt of its wings and tail. With a precision unsurpassed by any other bird of prey, the bird edges slowly forward . . . then stops! . . . edges forward . . . stops . . . all the while searching the ground below.

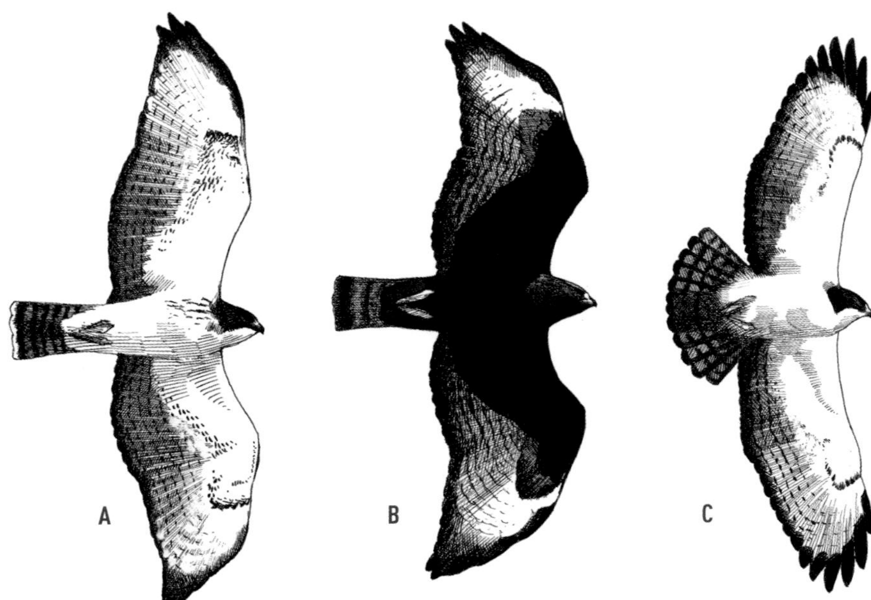

Short-tailed Hawk, underside. Light juvenile in shallow glide **(A)**. Dark adult in steep glide **(B)**. Light adult soaring **(C)**. Fairly small and stocky, reminiscent of Broad-winged Hawk but with more curved trailing edge of wings: pinched at base, bulging secondaries, long primaries often swept back. Light morph always very clean whitish on body and underwing coverts, contrasting with dark hood and dark gray secondaries. Dark morph distinguished from dark Broad-winged by shape and by dark secondaries and inner primaries blending to large pale area on outer primaries.

Short-tailed Hawk, upperside. Juvenile gliding **(A)**. Adult soaring **(B)**. Uniformly dark with faint pattern, unlike all other buteos, which generally show at least some pale mottling on back and uppertail coverts.

When the bird sights prey, it often parachutes vertically, slowly closing the gap, feet down, waiting for just the right moment to drop. At other times, it rockets down, spiraling in flight, entering the canopy at breakneck speed. If the attempt fails, the bird circles aloft and tries again. It seems that the only thing that tempers its tenacity is success.

This hunting behavior is virtually diagnostic. In Florida, the only other buteos to kite are the Red-tailed Hawk and the odd Swainson's Hawk. While the shapes of these two other buteos are somewhat similar, they are easily distinguished from Short-tailed Hawk by plumage and sheer aerial finesse. Next to the gravity-nullifying, stop-and-go control of Short-tailed, Red-tailed Hawk is an aerial amateur.

Short-taileds come in two colors, chocolate and vanilla. There are no intermediate plumages. In Florida, the dark morph outnumbers the light, but across Central and South America the opposite is true.

The adult dark morph is sooty brown on the body, upperwings, and underwing linings. It has the darkest, most uniform, most unpatterned plumage of any buteo. Its flight feathers are inconspicuously banded, imparting a grayish cast to the trailing edge of the wing; outer primaries, which are not banded, appear as a whitish patch near the tip of the wing. The contrast between the dark underwing lining and the gray flight feathers is stark; a dark trailing edge to the wing makes the pale flight feathers even more conspicuous.

The Short-tailed's tail is dark above, pale gray below. Like the wings, it is bordered by a dark subterminal band; it is also faintly and inconspicuously banded. The bird's namesake characteristic notwithstanding, the tail is not particularly short (by buteo standards at least).

Juvenile dark-morph birds look much like adults from above. Below, the dark body and underwing linings are finely spangled with small white spots. The dark border to the wings and tail is evident, but the tail pattern is overall somewhat less contrasting or distinct than on adults. From above, adult (and juvenile) light-morph Short-tailed Hawks are similar to dark morph (meaning uniformly dark brown). Below, the throat, body, and underwing linings are brilliant white. The gray flight feathers are similar to those of dark morph but, of course, on this bird, contrast markedly with the light morph's bright white underwing coverts. This wing pattern, a pale leading edge contrasting with a dark trailing edge, recalls light-morph Swainson's Hawk and White-tailed Hawk.

IN FLIGHT. The Short-tailed Hawk has long, wide wings. The wingtip is comparatively blunt. This wing shape easily distinguishes the Short-tailed from the long tapered-winged Swainson's Hawk and, with practice and study, Broad-winged Hawk. The wings of the similarly sized and proportioned Broad-winged are thinner and more pointed, clearly not blunt. On the Broad-winged, too, the trailing edge of the wing is pale with a dark border, not overall smoky gray.

Short-tailed Hawk soaring, approaching. Note strongly upswept wingtips.

A soaring Short-tailed is often confused with Red-tailed Hawk, another broad, fairly blocky-winged, short- and broad-tailed buteo that hunts aloft and habitually kites. The tail is a bit more squared off than Red-tailed's. In a full soar, the blunt wings of Short-tailed are thrust forward. The leading edge is clean lined and gently curved.

In a full soar, the Red-tailed also reaches forward, but mostly with the hand alone. The leading edge of the wing appears bulged or broken, not clean lined.

In a good wind, Short-tailed Hawk soars little but kites extensively. Sailing in lengthy glides that end in abrupt, gravity-defying stops, tacking back and forth like a quartering hound, these birds can even change altitude, rising vertically or descending slowly without overt effort or horizontal movement. The bird simply defies gravity.

In a glide, the silhouette of Short-tailed differs from that of Red-tailed. Particularly in a hard, fast glide, the outer wings are drawn way back. The bird seems to completely close its hand. The sharply tapered primaries slant back at a near 45-degree angle while the inner wing remains broad, unfolded, and fixed.

The resulting wing configuration seems contorted, all juts and angles. The bird's overall shape becomes squared off, blocky. This blocky configuration is particularly useful for distinguishing Short-tailed from Broad-winged, whose wings in a glide are smooth and gently curved along the leading edge, straight-cut along the trailing edge.

When folded up and gliding, the Short-tailed is fast, shooting across the sky with a speed that seems reckless for a buteo.

Head-on, Short-tailed has an unmistakable profile. It soars with flat wings, but with a distinctive upturn to the outer wings. When gliding, the flat-winged, raised-tip effect is also pronounced and it is something that Red-taileds simply do not show. Kiting birds, however, holding into a strong wind, draw their primaries way back. In this configuration, the birds may show a distinct dihedral.

Stiff, strong, but not particularly deep wingbeats characterize the Short-tailed's active flight. While stiff, the wingbeats are slower and less choppy than those of Broad-winged. Short-tailed's wingbeats are more suggestive of a larger buteo, a Red-tailed (or at least a Red-tailed on a smaller chassis).

Behavioral and structural similarities with Red-tailed Hawk notwithstanding, the bird that Short-tailed Hawk is most likely to be confused with is Broad-winged Hawk. Broad-wingeds and Short-taileds are about the same size and have similar overall proportions. In winter, in extreme southern Florida, the birds are found together, and in the Rio Grande Valley, during migration, a vagrant Short-tailed might easily be overlooked among the swarms of Broad-wingeds.

Adult and, particularly, juvenile Broad-winged Hawks appear pale below, but the body and underwing linings are not gleaming white, nor do they have a dark smoke-colored trailing edge to the wing. The wide white band on the tail of an adult Broad-winged is, of course, a dead giveaway, distinguishing it from the grayish-tailed, light- and dark-morph Short-taileds.

Dark-morph Short-tailed is most like juvenile dark-morph Broad-winged. Tail patterns are similar. Here, shape becomes a primary consideration. Short-tailed has a broader wing through the length, not the candle-flame shape of Broad-winged. In a soar or a glide, Broad-winged's wings are gently swept back, not sharply bent, angular, and overall blocky, as in Short-tailed. In profile, Broad-winged's wings are gently drooped or, in a harsh glide, sharply angled down. The wing configuration of Short-tailed is flat with gently raised tips.

A final and definitive point that relates to wing pattern: A dark Short-tailed shows wings that are darker overall than those of a dark-morph Broad-winged. The trailing edge is smoke gray, not silver-white. While dark Short-taileds do have paler wingtips, the white is confined to the outer flight feathers and projects itself as a whitish oval patch at the base of the outer primaries. On Broad-wingeds, the pale flight feathers extend along the entire wing.

Remember: A Short-tailed's secondaries are darker than the outer primaries. This contrast is true for both light and dark morphs (and just the opposite of adult White-tailed Hawk, which also differs from Short-tailed Hawk in shape and size).

The pale oval patch at the tip of Short-tailed's wing should separate Short-tailed from other dark-morph buteos.

A hunting Short-tailed that you can observe for any length of time is almost certain to distinguish itself by behavior. Many buteos hunt aloft, and a few kite. Only White-tailed Hawk approaches Short-tailed in precision stop-and-go control. No other buteo engages in the breakneck, sequential stoops that are the hallmark of Short-tailed Hawk.

1. An adult light-morph Short-tailed Hawk in a full soar. Bright white body and underwing linings contrast with darker flight feathers and a darkish (grayish) banded tail. The higher the bird, the stronger the contrast in the underwing pattern. Florida, January KK

2. An adult light-morph Short-tailed Hawk banking away. Note the touch of rufous on the neck. In shape, much like Broad-winged Hawk, but on Broad-winged, the trailing edge of the wing would not show an S-shaped curvature. This bird is less boldly patterned than the individual in #1. Florida, January MOB

3. A juvenile light-morph Short-tailed Hawk showing the classic soaring profile. Shape and patterning are fundamentally similar to those in adults. Overall, less contrasting in pattern than adults. Not quite so pointy winged as Broad-winged Hawk, and juvenile Broad-winged always shows some streaking below. Florida, February MOB

4. An adult light-morph Short-tailed Hawk. We know it is an adult based on tail pattern and the dark trailing edge on the wings (and the fact that this individual has been present for several years in the same location in Arizona), but the buffy or tawny tinge on the body and underwings is more typical of juveniles and very unusual for an adult. Arizona, January NH

5. An adult dark-morph Short-tailed Hawk showing classic pattern and shape. In Florida, dark morphs outnumber light morphs, but the ratio is reversed in the tropics. Florida, January KK

6. A dark-morph adult Short-tailed Hawk soaring and banking away. Banking turns are commonly tight and fast. Note the pale patches near wingtips (found on both light- and dark-morph birds). Compare this photo to dark-morph Broad-winged Hawk #7 and #9; pale area is more of a patch on outer wing on Short-tailed. Florida, January MOB

7. A dark-morph juvenile Short-tailed Hawk banking away. Juvenile dark morphs show mottled underparts that range, individual to individual, from spare white spots to light streaks. Florida, February MOB

8. A dark-morph juvenile Short-tailed Hawk. Same bird as in #7, now in a glide, showing mottled underparts and pale patches in the hand. In shape, somewhat like a Broad-winged Hawk, but with a longer hand and more curve or contour on the trailing edge of the wing. MOB

9. A dark-morph juvenile Short-tailed Hawk, soaring, wing-on. Note here the classic wingtips-swept-up, U-shaped configuration of the wing. No Broad-winged Hawk ever holds its wings this way. Florida, February MOB

Snail Kite

Like many of our regional specialties, the Snail Kite is not difficult to identify but can, at times, be difficult to find. Nonmigratory and geographically confined to southern Florida, the birds are nevertheless nomadic. Their peregrinations are an indirect response to changing water levels in the Florida wetlands they call home and a direct response to the effect water levels have on the birds' sole prey — the apple snail. Not enough water, and the snails decline and the birds must forage elsewhere. Too much water, and the snails lie too deep for the kites to reach. Again, the birds must move.

As abundant in the vast freshwater wetlands of Florida as Snail Kites once were, by the 1950s the United States population was perhaps as low as 50 individuals. Drainage efforts, development, and extensive flood-control measures all took a toll. Declared endangered in Florida in 1967, the species was afforded protection and has responded well to managed restoration efforts. Today there are about 1,000 Snail Kites in Florida.

Snail Kite aspirants can rely upon the assistance of hotlines, websites, list serves, nature centers, and fellow birders for information relating to water conditions and resulting concentrations of kites. Lake Kissimmee and the eastern end of the Tamiami Trail, near Shark Valley, are traditionally productive spots, but you are best advised to check the sources mentioned for current status. Snail Kites, like water levels, vary year to year.

This bird was formerly known as Everglades Kite, but that name did not reflect or do justice to the bird's extensive range in the Americas. In Florida, Snail Kites are found from the freshwater limits of the Everglades north to Orlando and particularly in the Kissimmee Valley and St. Johns River wetlands. They are fairly common in Cuba and locally numerous from central Veracruz south through Mexico and Central America. In South America, they are found south through Brazil and northern Argentina.

The Snail Kite is unlikely anywhere in the United States outside Florida, although there are five southern Texas records and one Georgia and one South Carolina record. Owing to a near total dependence upon apple snails, expansion seems unlikely.

IDENTIFICATION. Straightforward. Snail Kite is a medium-sized hawk, just a bit larger than Red-shouldered. Sexually dimorphic, the adult male is uniformly slate gray, with bright orange bill and legs.

Females are cold brown above, heavily and darkly streaked below. A bold facial pattern is manifest (but no more so than the distinct orange bill and legs). Juveniles are similar to females but overall warmer toned above, buffier on the face and below, and not so heavily streaked.

Wings of adult males are uniformly gray. The underwings of females and juveniles show a pattern similar to Northern Harrier, with secondaries darker than the primaries. On females and juveniles (rarely adult males), there are also pale translucent patches at the bases of the primaries (somewhat like Red-shouldered Hawks).

The triangular tail is white at the base; the distal half is black, narrowly tipped

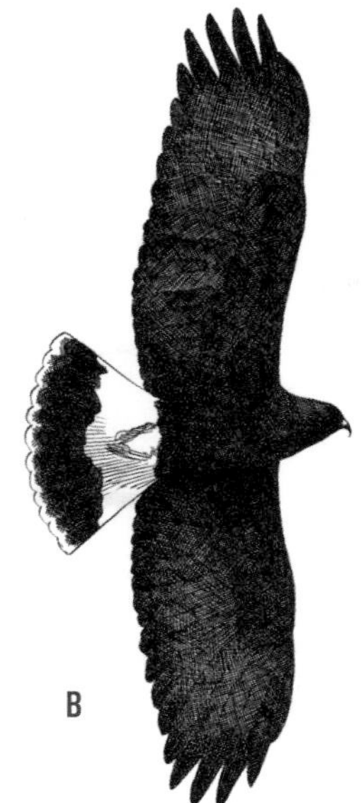

Snail Kite, underside. Juvenile gliding **(A)**. Adult male soaring **(B)**. Distinctive. Broad cupped wings with long primaries; relatively short square tail with large white area at base. Shape of soaring and gliding birds most similar to Osprey, but broader winged, with more floppy actions, and very different plumage. Adult male uniformly slate gray; adult female and juvenile brownish, streaked, with pale area on outer primaries and bold eye line. Adult female more heavily marked than juvenile, with more sharply defined tail band.

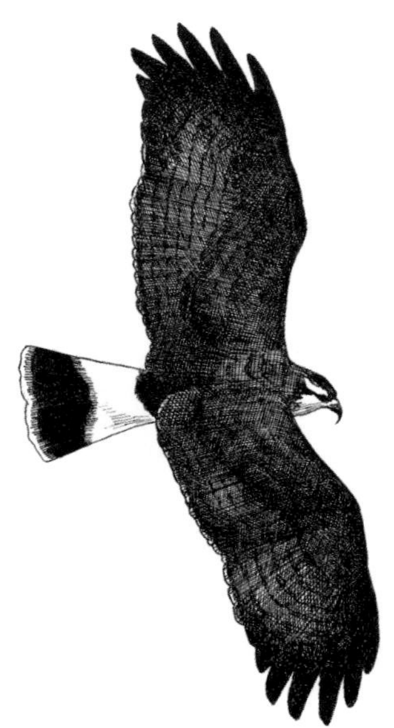

Snail Kite, upperside, juvenile. Very broad wings. Generally dark brown with white on rump and base of tail, blackish tail band, white tip.

with white. All plumages have this distinctive tail pattern, which recalls (should you be familiar with it) a Wilson's Storm-Petrel.

The Snail Kite has long thin legs and a highly exaggerated, long, thin, sharply hooked bill that it uses to extract snails from their shells. This bill is so distinctive that it serves as a field mark in and of itself.

Snail Kites are often seen carrying snails to feeding perches — sometimes the snail is carried in the bill, sometimes in the bird's talons. Snail Kites are highly gregarious. If you find one, you will probably find a group — eight, ten, or more.

IN FLIGHT. The Snail Kite appears very buteolike. Wings are long, but exceedingly broad, recalling an eagle or Common Black-Hawk. The label *paddle-shaped* applies, but the wing of Hook-billed Kite conforms more to this shape. Snail Kite's wings are more squared off and boxy.

Snail Kite, adult male (left) and adult female (right), searching for and dropping on snails.

The primaries of Snail Kite are long and flexible. The tail is relatively short, triangular, and straight-cut when fanned and, as might be expected from a kite, is often in use, rotating and swiveling continuously to balance the bird in flight. When the bird is gliding, with tail closed, the tip is slightly forked.

In a full soar, the wings are thrust forward with a slight Red-tailed-esque bulge at the wrist. When flapping or gliding, the outer wing is swept back. Active flight is floppy but purposeful, with a pumping cadence. In hunting flight, the bird pumps along four to six flaps in a sequence, then executes a short glide. While hunting, the bird's head is lowered as it scans for prey.

This low searching flight, coupled with the conspicuous white uppertail and the bird's tendency to droop its legs, easily evokes the image of Northern Harrier. But Snail Kites fly with wings severely bowed down along their length, not uplifted in a dihedral, and Snail Kites do not rock in flight as harriers do. Also, the head and neck of Snail Kites are more prominent.

Graceful when soaring and nimble, even acrobatic, in its hunting, Snail Kite may not look like the slimmer-winged kite species, but in flight its affinities are manifest. The bird's overly broad wings enable it to glide just above stall speed. It is not actually kiting, nor is it hovering (although glides are interspaced with a few flaps). It seems as though gravity is more lenient with the Snail Kite than it is with other birds. While most other birds glide, Snail Kite floats.

The Snail Kite often punctuates lengthy glides with bouts of soaring (often in tight circles), then resumes gliding. When circling, birds are probably hunting, not necessarily seeking lift. It is not uncommon for other Snail Kites to fly in and join soaring birds.

Head-on, the Snail Kite shows *extremely* bowed wings, raised at the base and drooped at the tip. The resulting graceful arch curves the entire length of the wing, like a Red-shouldered Hawk (but more so). Of all the raptors, only Osprey has

Adult male Snail Kite gliding. Note cupped wings and long flexible primaries.

more bow in the wings, but the configuration on Osprey is angular, jutting up, then angling down. On Snail Kite, the curvature is fluid and seamless.

Snail Kite both flaps and glides on bowed wings. In a full soar the bowed configuration is less obvious yet still evident. This cupped-wing configuration is its signature field mark.

As might be expected with a bird as broad winged as Snail Kite, the flap is heavy. Others have described it as "floppy," which fairly describes the downstroke, but the upstroke is quick, snappy, somewhat like Sandhill Crane's. Much of the flap is above the body (like Bald Eagle's).

Perhaps because Snail Kites are so social, a display flight is frequently seen — a stiff, exaggerated flap with deep, slow wingbeats executed at medium heights. The cadence of the flaps and the severe arch of the wings are similar to the flapping flight of Magnificent Frigatebird.

In the final analysis, there are only a few species that might be confused with Snail Kite. Harris's Hawk has a similar shape and, at a distance and in poor light, shows a similar plumage pattern. But even if Harris's Hawk were found in Florida (which it is not), the parabuteo's chestnut wing coverts and wing linings (and broad, not narrow, white tip to the tail) should easily distinguish it.

Northern Harrier has a white rump (but not a half-white tail!) and, like Snail Kite, hunts low and slow. But a harrier flies with a marked dihedral; Snail Kite flies with wings deeply cupped or bowed.

A vagrant Snail Kite in southern Texas would need to be separated from Hook-billed Kite, a bird with which it shares much structural similarity. A male Hook-billed Kite might be gray above but would have barred underparts and a longer, boldly banded tail.

In the end, there is little with which to confuse a Snail Kite.

The Snail Kite is one of the planet's most specialized raptors. Seeing one requires a special effort. But to see Snail Kites circling above their Florida stronghold is not only to enjoy one of our most distinctive raptors, but to experience one of our most celebrated ecosystems. The Everglades may not be the endless sea of grass it once was, but the sight of soaring Snail Kites may make it seem that time, like gravity, has been suspended for as long as the birds remain in view.

1. An adult male Snail Kite carrying nesting material and banking slightly. Note the blocky, broad wings and triangular tail; also note the slaty gray plumage and white rump. At this range, the slender, shell-picking bill (framed in orange) is an eye-catching (not to mention unique) characteristic. Florida, February KK

2. An adult male Snail Kite in a glide. The acutely bowed dihedral is very unlike the raised dihedral of Northern Harrier, another raptor that might be seen gliding low and slow over the Everglades. Florida, February KK

3. An adult female Snail Kite. A brown raptor with bowed wings, streaked underparts, and a squared-off white base on the broad angular tail, and it's carrying an apple snail. This combination doesn't leave much to the imagination. Florida, May NH

4. A nearly one-year-old female Snail Kite in direct flight, carrying nesting material. Take note of the silvery patch on the outer wings below and the white at the base of the tail. While photos such as this might suggest a resemblance to the Northern Harrier, all similarity disappears when you watch a Snail Kite in motion. Florida, February KK

5. The dorsal view of a juvenile Snail Kite, banking. Note the paddle-shaped wings thrust forward (typical of soaring birds) and the subtle banding on the flight feathers (typical of juveniles). Juveniles show a warmer brown plumage than adult females. The white at the base of the tail is superficially similar to harrier, but much else is not. Florida, April CS

6. An adult male Snail Kite hunting. Low, slow, and buoyant flight broken up with bouts of floppy-winged hovering: that's how Snail Kites move across the marsh. Note the acutely bowed wings and the twisted tail, indicative of a kite maneuvering in the air. Florida, April CS

7. An adult male Snail Kite soaring. Paddle-shaped wings reaching forward and white triangle on the tail displayed. The shape is similar to Hook-billed Kite, but the wings are longer and there's considerably less pinch at the base of the wings on Snail Kite (and, of course, the tail pattern is very different). Florida, January KK

8. An adult male Snail Kite soaring, with a second bird gliding above. Compare the soaring bird's shape to Harris's Hawk #7, another paddle-winged raptor. Similar? Yes. But the configuration of Snail Kite's wing is unique. (Plus, when making this choice, one of the two would be well out of range.) Notice also the acute gliding profile in the higher bird. Florida, April CS

9. A juvenile Snail Kite in a high banking turn, showing wing molt as it nears one year old. This young bird shows the same basic brown-and-streaked-underparts pattern of adult females, but is overall paler and with a less crisply defined dark tail band. Florida, April CS

Regional Specialties

Raptors of the Rim

SPECIES
Hook-billed Kite, *Chondrohierax uncinatus*
Aplomado Falcon, *Falco femoralis*
California Condor, *Gymnogyps californianus*

The raptor enthusiast eventually reaches the point where there are few North American raptors left to see. But in life (as well as in this book) the best is often left for last. North America's remaining three raptors — the Hook-billed Kite, Aplomado Falcon, and California Condor — are species so geographically restricted that hopeful observers will need to make a special pilgrimage (and perhaps apply special effort) to find them.

Not long ago, seeing any of these regional specialties was unlikely, even impossible. To a varying degree, all owe their existence to concerted management (if not outright reintroduction) efforts.

The Hook-billed Kite, a bird of tropical forests, was first found nesting in the United States near Alamo in the Rio Grande Valley in 1964. At the time, habitat loss and alteration in the region threatened the populations of many borderland bird species and undermined hopes that Hook-billed Kite would become permanently established. Fortunately, in the final two decades of the twentieth century, efforts by the U.S. Fish and Wildlife Service to protect the Rio Grande corridor gained momentum and, more important, ground — habitat! Today, the Lower Rio Grande Valley National Wildlife Refuge and satellite parks and preserves serve as a secure biological foundation for the current United States population of approximately 20 pairs of Hook-billed Kites — although Homeland Security concerns threaten to undermine this ecological gem and the future of Hook-billed Kite in the United States.

The Aplomado Falcon, a colorful desert grassland falcon, was extirpated from the United States by the 1950s. The last known nesting attempt in Texas occurred

LEFT: This is truly a raptor of the rim: feet down, slotted primaries up, an adult California Condor rides an updraft along the South Rim of the Grand Canyon. CS

An adult Aplomado Falcon at rest along the Texas coast. Okay, we know it's not in flight, but being an Aplomado, it soon will be. NH

in 1941; in New Mexico, 1952. Beginning in 1985, efforts to restore the bird to its former United States haunts were initiated. Today, as a result of captive breeding, release, and, more recently, natural nesting, Aplomado Falcons may again be seen in several places. Biologists express guarded optimism for the birds' future in the United States.

Unlike the previous two species, whose broader world populations were always secure, the California Condor's demise in the United States meant total extinction. Restoration efforts were not simply a matter of restoring a bird to its former range or encouraging a bird to expand its range or even securing a future for the bird in the wild. It was a matter of saving one of the planet's oldest and most celebrated creatures. As a result of a focused breeding program and a determined reintroduction effort, California Condors are once again flying and reproducing in the wild.

The last three species accounts focus on putting these raptors of the rim within reach of your skills.

Hook-billed Kite

The Hook-billed Kite is an improbable bird, having both an unusual shape and unconventional habits. It can be hard to find and, when found, hard to see. But when you do finally see the bird soaring over Bentsen–Rio Grande Valley State Park or Santa Ana National Wildlife Refuge, you will experience one of the true treats of raptor watching in North America.

Hook-billed is a tropical kite. Dissimilar shapes notwithstanding, it is most closely related to Swallow-tailed Kite. Hook-billed reaches the northern limit of its range in southern Texas and has never been seen north of the Rio Grande Valley corridor. There are approximately 20 nesting pairs in the United States, all found between Brownsville and Falcon Dam. This makes Hook-billed Kite the rarest nesting raptor in the United States.

South of the United States border, Hook-billed Kites are widespread and common throughout Mexico and Central and South America, south to Brazil. Because of their very specialized needs, low density is the rule in many areas.

Hook-billeds are migratory. One of the many discoveries resulting from fall migration counts at Veracruz, Mexico, is that some Hook-billeds vacate the northern part of their range. While the birds are not to be expected at hawk watches north of their established range, farther south, at Cardel in central Veracruz, up to 180 in a season have been counted moving south.

The Hook-billed is a forest hawk, primarily a bird of the understory and lower canopy. It is found most often in riparian and tropical deciduous woodlands but also in rainforest and cloud forest. The bird's presence or absence is predicated upon an abundance of tree snails. So specialized is the bird's diet that, over its range, bill size is directly related to the type and size of tree snail found there.

As might be expected, given the less than elusive quality of its prey, the Hook-billed Kite is not a dashing hunter. It is, without question, the most unraptorlike raptor covered in this book. Even the Snail Kite, another snail-obligate species, makes a show of hunting.

When perched, the Hook-billed resembles a bowling pin fitted with a goofy, parrotlike bill. When foraging, the bird's movements are slow, methodical, even slothful. To secure prey, the Hook-billed more or less falls upon the prey or lofts itself toward it (the effort hardly qualifies as "flight"). The bird plucks the snail from a tree trunk or branch. An even less energetic (but often used) technique involves walking along a limb to secure prey.

For many birders, seeing the Hook-billed Kite has meant plodding the trails of Santa Ana National Wildlife Refuge (sometimes for days), and many who have depended upon this technique have left the valley kiteless. There is another way to find the bird, one more in keeping with the spirit of hawk watching.

Despite its generally sedentary lifestyle, the Hook-billed Kite is an accomplished flier. Though reported not to soar a lot, the bird usually takes to the air in midmorning and again in late afternoon; when courting (in April), Hook-billed Kite may spend considerable time in the air. Good vantage points for seeing the birds include the flood dikes at Santa Ana National Wildlife Refuge and Bentsen–Rio Grande Valley State Park.

IDENTIFICATION. Hook-billed Kite is a medium-sized raptor boasting an accipiter-length tail and outlandishly large bill. Adult males are overall gray, resembling, somewhat, a dark Gray Hawk. Females are brown backed and richly barred with rufous below, and they have orange cheeks. Juveniles are like females but are mostly white below (overlaid with spare orange-brown barring) and have white cheeks.

All have boldly banded tails. From below, the adult male has two wide white bands on a dark tail (bands appear gray from above). The female's tail bands are light brown on dark brown. The juvenile has multiple bands — its tail looks much like the tail of juvenile Cooper's Hawk.

Hook-billed Kite also has a dark morph. To date, a single dark-morph bird that has taken up residence in Bentsen State Park is the only one recorded in the United States. The dark-morph adult male is sooty black without any barring and with only a single broad white band on the tail (not two). The female and juvenile dark morphs appear mostly brownish black.

In flight, the white tail bands of the male contrast with the slate gray body, are visible at some distance, and help identify the bird. If well lit, the gleaming white undertail coverts combine with the inner tail band to create a white "patch" effect that may be transmuted by distance and incorrectly ascribed to the rump (not the undertail) of the bird. This same optical illusion causes distant, immature Cooper's Hawks to sometimes be mistaken for Northern Harriers.

Hook-billeds have a dramatic wing pattern. The primaries are boldly barred or spotted, embedded with an array of tiny windows that allow light to shine through, making the entire hand seem translucent. When a Hook-billed is backlit, these glowing outer wing panels are visible at extreme distances and make an exceptional field mark.

The translucent wingtips are most prominent on adult females. On females, pale patches are white near the tip and become rufous along the inner primaries. Adult males are variable but generally show more patterned spotting in the primaries, somewhat muting the translucent patch found on the female.

Juveniles show the least translucence in the outer wing. What light does transfuse through resembles the crescent window pattern seen on the wing of Red-shouldered Hawk. While Hook-billed Kite and Red-shouldered Hawk do overlap in range, and do share some structural similarities (both are, after all, forest birds), structural *dissimilarities* easily distinguish them.

IN FLIGHT. Hook-billed Kite shows one of the most distinctive shapes in all raptordom. In fact, the overall shape of the bird is Hook-billed Kite's most distinctive field mark. Nothing else that soars over North American skies looks quite like it.

The body seems slender and slight, too narrow to support the wings. The head is long and projecting. The tail is likewise very long, accipiter-length, and appears too narrow for the rest of the bird when closed but wider and better proportioned when partially fanned (which is as fanned as it ever gets).

The wing configuration is distinctive, bordering on unique. Disproportionately broad, the hand appears every bit as wide as the arm and wider. Pinched along the trailing edge (and narrowest at the base) the slender-bodied bird seems fitted with big, broad, paddle-shaped wings — and not canoe paddles. We're talking near ping-pong-class paddles.

Hook-billed Kite, underside. Juvenile gliding **(A)**. Adult female in shallow glide **(B)**. Adult male soaring **(C)**. Wings are broad tipped and paddle-shaped in the extreme, actually narrower at base than at inner primaries. Wings pushed forward in soar and in glide; primaries seem to be held perpendicular to body. Tail square tipped but bulges on sides. Adult male has slightly broader wings than adult female, which has slightly broader wings than juvenile, but even juvenile is strikingly shaped. Adult male is darkest, with mostly gray plumage including all secondaries and broad gray bars on primaries. Adult female has rusty bars on body and underwing coverts, narrow blackish bars on primaries and secondaries. Juvenile is variable but usually pale on body and underwing coverts with fine brown barring, narrowest dark bars on primaries and secondaries, and more dark tail bands than adults. All show massive bill and pale green loral spot, which can be surprisingly visible.

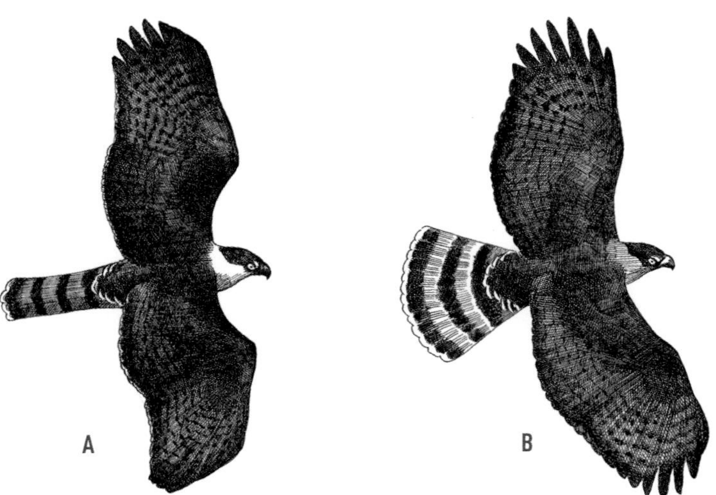

Hook-billed Kite, upperside. Juvenile gliding **(A)**. Adult female soaring **(B)**. Fairly uniformly dark except gray tail bands and pale collar, which is rufous in adult female and whitish in juvenile (dark gray in adult male). Adult female has some rufous on inner primaries visible from above.

Other North American raptors have a similar paddle-winged configuration — prominent among them Harris's Hawk and Snail Kite. But no other North American raptor's wings are as broadly ovate as Hook-billed Kite's.

The paddle-shaped configuration is most apparent when birds are soaring with wings thrust slightly forward and wingtips turned slightly back. When they glide, the paddle-winged effect is diminished somewhat. Also of note, the configuration is less pronounced in juveniles than in adults. Nevertheless, at any age, the breadth and shape of Hook-billed Kite's wings are manifest. In combination with the narrow projecting head and long narrow tail, it is unique.

Hook-billed Kite gliding head-on (top), and soaring head-on (center); adult male soaring in approach (bottom). Distinctive wing shapes in all attitudes. Soaring with wings pressed down but tips bent up produces the effect of wings thrown well forward when seen from certain angles. Gliding birds set wings sharply down, often *whiffling* or sideslipping to lose altitude quickly when approaching trees.

Soaring and gliding are accompanied by little or no flapping. Flapping is normally used to gain altitude (or reach thermals) quickly. The species sometimes kites briefly but never (to the best of our knowledge) hovers.

The way Hook-billed Kite holds its wings when soaring is almost as diagnostic as the wing shape. Soaring birds hold their wings angled down along the arm, uplifted along the hand, creating a reversed bowed effect. (If you are familiar with the configuration of the World War II Corsair fighter plane, you can picture the bird.) When gliding, wings are gently down-curved along their length.

When soaring, Hook-billeds often seem to turn flatly; that is, they don't bank at an angle to make the turn, like other hawks do. Think of the flat turns that a Black Vulture makes. Actually, a distant male (dark) Hook-billed Kite can look like the squat-winged Black Vulture — except for the long tail.

Long it may be, though it may not always be obvious. Hook-billed Kite commonly soars with the tail closed, with diminished surface area, making the tail difficult to detect well short of the point where the bird's wide wings become muted by distance. It makes sense, actually. With wings so broad and Texas thermals so strong, who needs an open tail?

For most of the year, Hook-billeds do not spend lavish amounts of time in the air. They commute between roost sites and feeding areas in the morning and afternoon. During the nesting season, they may be surprisingly active, soaring high, covering a lot of sky easily and quickly. They will stoop fast and steep, not to encounter prey but to regain the canopy quickly, and they will *whiffle* (spill air from the wings by sideslipping), the way incoming geese will do. While Black Vultures do this, other raptors seem not to.

The flap is languid, stiff, slow, heavy, and shallow. Thanks to the width of the wings, there's a lot of air to move, a lot of resistance. The arc of the wing is centered mostly above the body, and while it may at times appear choppy, it is a slow chop. The wingbeat has a measured cadence to it — similar to the metronome-like pumping flap of a Great Blue Heron.

In spring, Texas Hook-billed Kites are often seen displaying high overhead. When displaying, they exhibit a deeper, more exaggerated flap. Juveniles have also been seen engaging in this display flight.

Because of its unique shape, Hook-billed Kite is not often mistaken for other species. A Hook-billed Kite soaring with migrating Broad-wingeds (as they sometimes do along the Rio Grande Valley) is as distinctive as a pickup truck parked among a fleet of mid-sized sedans — about the same size but more bulk. At a distance, a dark male Hook-billed has a broad enough wing to appear eaglelike, and distance can mute the shape of the paddle wings. The longer, accipiter-like tail of Hook-billed should help distinguish kite from eagle at all but extreme distances, and Hook-billed Kite is also conspicuously smaller than an eagle (so shows a smaller bird's GISS). Also, eagles are rare in the Rio Grande Valley.

Gray Hawk has a plumage that is similar to male Hook-billed Kite's, but a buteo shape, not a bowling pin with paddle wings. A soaring Snail Kite (normally not in the Hook-billed Kite's range) will show cupped or down-bowed wings. Hook-billed shows just the opposite, an inverse bow — arm down, hand up.

Harris's Hawk, with its long tail and paddlelike wings, is perhaps most like Hook-billed Kite in structure, but the dramatic tail pattern on Harris's Hawk is evident at great distances, and there is a better-than-even chance that the chestnut wing coverts and wing lining will show as well.

Juvenile Hook-billeds, showing a diminished paddle-wing shape and less translucent wings than adults, may be mistaken for wide-winged, long-tailed buteos — in fact, they can be mistaken for juvenile Red-shouldered Hawks. But while the similarities between these two forest buteos may be structurally fundamental, the differences are vast, a list that merely begins with a larger wing patch on the kite, a different flap, a different set to the wing, and a distinctly narrower tail.

As a first and last resort, *just look at the overall shape of the bird.* It truly is the best field mark.

1. A card-carrying (or snail-carrying) adult male Hook-billed Kite in powered flight. Note large bill, slender tubular body, and long tail. Most of all, note the enormous paddle-shaped wings. Also uniformly gray plumage (males), banded tail, and barred wingtips. Texas, July WC

2. An adult female Hook-billed Kite in powered flight, carrying a tree snail. Orange cheek and boldly barred underparts easily distinguish females from males. Juveniles are similar to females but more lightly marked and show a white cheek. Note the pale crescents at the wingtips. More than one observer has confused this bird with Red-shouldered Hawk. Texas, July WC

3. An adult female Hook-billed Kite flapping; same bird as #2. Besides paddle-shaped wings and small head, note coarse orange barring on the body, boldly barred flight feathers, and hints of rufous in the outer primaries. WC

4. A likely subadult male Hook-billed Kite in full soar. The shape *is* the field mark. In Texas, it might be confused only with Harris's Hawk and, at first glance, Red-shouldered. Note, in particular, the paddle-shaped wings that are acutely pinched at the base; and also the tail, which, though barely fanned, is about as wide as the tail commonly gets. Texas, February KK

5. An adult male Hook-billed Kite soaring; this is a particularly dark bird. One true dark morph (wholly dark except for a single broad white tail band) has been recorded north of Mexico. Texas, March TL

6. A soaring adult female Hook-billed Kite. Backlit silhouette, showing translucent barred wingtips and distinctive paddle-shaped wings. Left wing is very obviously damaged. Texas, March TL

Aplomado Falcon

From the coastal prairies of Texas, west through the Trans-Pecos, to the high desert grasslands of New Mexico and southwestern Arizona, the striking Aplomado Falcon was once a common sight. But around the turn of the twentieth century, this grassland falcon began a decades-long decline that would end with its extirpation from the United States. The last known breeding effort, prior to reintroduction effort, was in New Mexico in 1952. This falcon, then, has the distinction of being the only native hawk to have been extirpated from the United States.

In the ensuing decades, there were scattered sightings of Aplomados in the borderlands, presumably wanderers from south of the border. Beginning in 1985, a reintroduction program was initiated in southern Texas. It was successful, and in 2006 it was expanded into southern New Mexico. An estimated 40 pairs of Aplomado Falcons are now breeding north of the border (mostly on the lower Texas coast). Laguna Atascosa National Wildlife Refuge is the center of the release program and the core of the birds' United States range. Released birds are breeding successfully, and birds from the wild native population in Mexico have also been drawn to the United States side of the border. Biologists are optimistic about the bird's future, and raptor enthusiasts are elated that this beautiful, exciting falcon may once again be a key component of the southwestern grassland biota.

South of the United States, the Aplomado remains fairly common. It is numerous from Veracruz, Mexico, southward and found in a variety of open habitats throughout South America (giving the bird one of the widest resident ranges of any New World raptor).

Aplomados are considered nonmigratory. Paired adults remain together throughout the year, and young stay with adults for a protracted period after fledging. Cooperative hunting is a common and characteristic behavior. One bird will attempt to flush prey while a second waits to give chase. They are active, dashing hunters of birds and will take prey as large as chachalaca and as small as Yellow Warbler. They have been known to capture bats after sunset (proving what agile fliers Aplomados are).

At times, they are insectivorous, feeding on moths, beetles, dragonflies, grasshoppers, and locusts. They follow grass fires, hovering in wait for flame-flushed prey. Aplomados are also kleptoparasitic, adept at stealing prey from other birds of prey (and given the ferocity and tenacity of these pursuits, it is not at all certain whether the Aplomado is more interested in the prey or its predator).

The name *aplomado* is open to interpretation. It is most often construed to mean "lead-colored," a presumed reference to the deep blue-black color of the adult's back. It may also be a derivative of *plumb* or *plumb bob,* a vertically hanging lead weight, suggestive of the plummeting stoop of a falcon. Given that twisting chases are more in the character of the bird than vertical stoops, the first interpretation seems more apt.

IDENTIFICATION. Aplomado Falcon is a medium-sized falcon — slighter and modestly smaller than Peregrine, proportionally more like American Kestrel. Think of

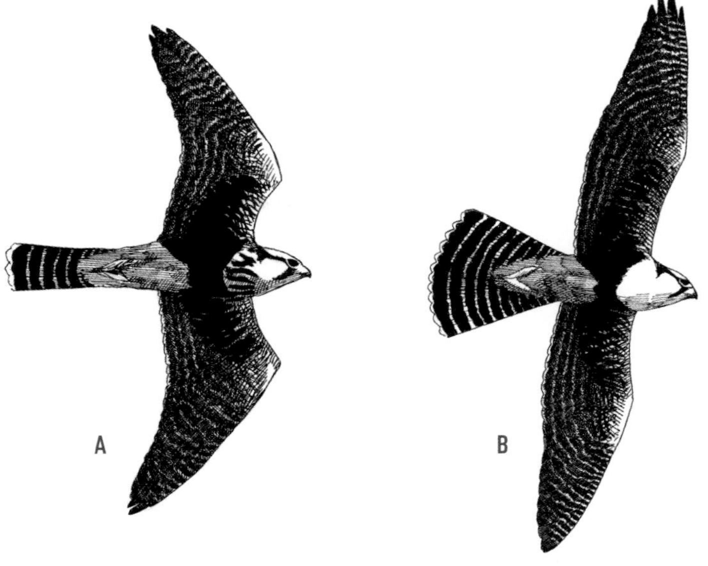

Aplomado Falcon, underside. Juvenile gliding **(A)**. Adult soaring **(B)**. A strikingly long-tailed and long-winged falcon, always shows black belly band, dark underwings and tail with fine white bars, slate gray upperside, and neat face pattern.

Aplomado Falcon, upperside. Adult gliding. Uniformly dark slate gray above with very narrow white bands on tail, bold white eyebrow.

it as a large, long-tailed American Kestrel and you've got the picture. Compared to other falcons, Aplomados appear small headed. This appearance holds true whether the birds are perched or in flight.

Aplomado Falcon is colorful and dramatically patterned, and both adult male and female are similar. The adult's slate blue back is complemented by a black cummerbund, a belly band reminiscent of a juvenile Rough-legged Hawk's. The chest is creamy white; the lower belly and undertail coverts are cinnamon. The head is boldly patterned: Aplomado wears a dark cap separated from a prominent dark eye line by a white supercilium, and it sports a short-trimmed mustache.

In sum, the birds' colorful contrasting patterns are visible at a great distance and greatly facilitate identification.

In flight, the black belly band bleeds out across the underwing coverts, ending with a subtle tawny carpal patch. Also from below, the flight feathers are overall paler (at close range, finely spangled), and the long dark tail is divided by a concentric array of pencil-fine white bands. Some birds may lack the bands and show an all-dark tail.

Juvenile Aplomados are like adults except for a browner back, streaked chest, and paler, buffy (not cinnamon) belly and undertail coverts. While many guides depict adults with a white chest and juveniles with a buffy chest, expect variation. The pale breast on adults varies from white to cream. The undertail coverts may range from rufous to cinnamon to tawny.

IN FLIGHT. The Aplomado Falcon is an elegantly slim falcon with long narrow wings and an exceedingly long tail. In shape and manner of flight it resembles an overgrown American Kestrel.

Like kestrel, Aplomado's wings are short in the arm, long in the hand, and not severely tapered. Wingtips, in a soar, are quite blunt, somewhat akin to Prairie Falcon's.

The tail is narrow and untapered (or slightly flared); the tip, slightly rounded. In active flight or when the bird is gliding, the tail is closed, accentuating its already considerable length. Distant or high-soaring birds appear headless, all wing and tail.

An Aplomado's wingbeats are stiff and fluttering, the flap more kestrel-like than the whippy, elastic flight of a Peregrine. But Aplomado is much larger than an American Kestrel, so the flap is slower and projects greater power, more oomph. In unhurried transit, Aplomados fly with a somewhat rowing wing motion, with a snappy, steady cadence.

Aplomados are active, energetic fliers, flapping a great deal with many beats leading up to a lengthy glide. When going into a glide, they seem to *snap* the wings into locked position — short arm thrust forward, long hand pulled back, wrist sharply bent.

Most glides are fairly shallow, not at all like the steep, fast, closed-winged, hard-angled, altitude-sacrificing glides of Merlin and Peregrine. The species appears extremely buoyant when gliding, the aerodynamic advantage afforded birds with long narrow gliderlike wings and tail. When gliding, Aplomado Falcon holds its wings bowed slightly down along their length, once again suggesting American Kestrel. The birds soar on flat or slightly decurved wings.

Actually, the birds do very little soaring, preferring to tack back and forth in flight. You are far more likely to find Aplomado Falcon in active flight, at low to medium heights, than circling high overhead.

Backlit, showing no plumage pattern, a distant soaring Aplomado can look much like a Mississippi Kite. This similarity is heightened by the Aplomado's pale trailing edge on the secondaries, a trait it shares with Mississippi Kite. The translucent wing pattern, however, is more exaggerated on the falcon — in fact, a backlit Aplomado will show a translucent wing pattern reminiscent of Black-legged Kittiwake.

The hunting behavior of Aplomado is distinctive enough to be considered a field mark. An alert and agile predator, it hunts from a perch but seems loath to stay there long, repeatedly flying out to intercept potential prey or to engage other Aplomados or hawks.

It also hunts on the wing, patrolling low over the beach, dunes, or fields, tacking back and forth like a hound trying to pick up a scent. Flying high, it may suddenly dive in a closed-winged stoop, pull out, then rocket upward, only to execute a smart wing-over and stoop again in a fast, vertical, wing-pumping plummet. It's as if the bird were saying: "You think I'm a kestrel? Well watch this!"

Departures from perches are quick and dramatic, and while the Aplomado's flight never really projects the sense of speed and power of the larger falcons, it is nevertheless adequate to the task — or maybe just more accomplished. As nimble as the bird needs to be, its movements seem almost to have a perfunctory quality, a sense sometimes that the falcon is only half trying and that there is plenty of flash and dash in reserve (if the bird ever needs it).

Still, the bird is a character shifter — behaving much like a small falcon, then *wham,* shifting into overdrive and acting like one of the larger falcons. But in the end, there is little chance of mistaking Aplomado Falcon for another hawk — including any other falcon. This species is distinctive enough in shape, plumage, and mannerisms to set it apart.

If a bird is backlit so no pattern is evident, an Aplomado might be mistaken for a Mississippi Kite. Both are agile fliers, but the kite would be buoyant and graceful, its proportionately shorter tail noticeably flared out at the tip. Its wings, too, would be broader.

Among falcons, the Aplomado is most like an American Kestrel — an overgrown kestrel with an outsized tail and a measure of dash in its demeanor. Watching both birds in the sky together, you will not fail to note the similarities. But similar is not same. Size and weight separate the two birds, and these differences are projected in how each species flies. To envision Aplomado, think of a kestrel the size of a Peregrine or a Prairie Falcon with a similar verve. Apt as this description might be, it is not the same as seeing the bird in the flesh.

1. An adult Aplomado Falcon. One of the most richly patterned of the falcons. All dark above, body tricolored below (white throat; breast and buffy orange undertail coverts separated by a dark belly band). Note the white trailing edge of the wing. A nervous, active species, it seems loath to perch too long. Notice the color band. Aplomados are being reintroduced into their former range. Texas, October NH

2. An adult Aplomado Falcon. Note the overall kestrel-like slenderness of the bird. Pale throat and boldly patterned head contrast with overall dark underparts. Active flight is the bird's MO. Panama, August MOB

3. A juvenile Aplomado Falcon in fast, powered flight. Except for streaking on the breast and paler undertail coverts, the plumage pattern is similar to that of adult. Argentina, November TJ

4. A juvenile Aplomado Falcon in casual flight. The Aplomado's long slender wings and exceedingly long narrow tail have caused more than one observer to comment on how very kestrel-like this bird may appear. Veracruz, October CS

5. A juvenile Aplomado Falcon flapping and banking away. Pale areas below are somewhat variable, from cream- to rust-colored. Texas, July WC

6. A juvenile Aplomado Falcon in a glide. Notice how the wing shape and long narrow tail easily conjure the image of American Kestrel. But note the bold head pattern and tricolored underparts (pale breast, dark belly, pale undertail coverts). Veracruz, October CS

7. A juvenile Aplomado Falcon in fast direct flight. This is a particularly rusty-colored individual. In proportion, the tail is longer than any other falcon. Veracruz, October. CS

California Condor

In the mid-1980s, when the first edition of *Hawks in Flight* was written, we never anticipated having the fortune and privilege of including this bird in any future edition — except, perhaps, as a poignant footnote. The condor was that close to extinction.

Ten thousand years ago, during an age when mastodons and camels ranged across much of North America, California Condors were widespread, occurring in New York, the Carolinas, and Texas and along the Pacific Coast. When Europeans reached the continent, they found the bird only along the Pacific Coast, the population maintained first by a surfeit of large deceased marine mammals and later by cattle brought by the Spanish.

In 1805, Lewis and Clark tried to shoot a condor on the Columbia River in Washington State. This effort was unsuccessful but, from the standpoint of human-condor relations, prophetic. By the 1930s, because of wanton shooting, secondary lead poisoning, scientific collecting, and loss of habitat, the remaining California Condors on the planet were making a last stand in the mountains north of Los Angeles, California.

It was a battle they lost. In 1986, the last free-flying California Condor was taken from the wild in an effort to keep the bird from extinction.

Today, as a consequence of a successful captive-breeding program and intensive reintroduction efforts, free-flying California Condors once again spread their wings across parts of California, Baja California, Arizona, and southern Utah. From a remnant core of 27 birds in 1986, just under 400 condors were alive in 2011 — 109 in California, 19 in Baja, Mexico, and 70 in Arizona (principally in the vicinity of the Grand Canyon), with 201 birds held in captivity. California birds range from the hills above Los Angeles north to Pinnacles National Monument (where they have recently bred). Grand Canyon birds regularly travel to Zion National Park in southern Utah, principally in the summer months.

The California Condor is not just a bird. It is a symbol — of the plight of native wildlife in the face of an exploding human population and all its associated pressures, and of the reparation our species can bring about when we dedicate ourselves and our resources toward it. Biologists are optimistic about the future of the bird but warn that it may never outgrow the need for active management and protection.

Nevertheless, condors are back in the wild, and students of raptors can see them — the largest and grandest bird that soars in North American skies.

IDENTIFICATION. Key to California Condor identification is geography. The birds are so far highly localized, found primarily where captive-bred birds have been released. Outside of these reintroduction sites, observers have severely diminished chances of seeing condors.

While nonmigratory, California Condors engage in seasonal movements through their ranges — and their ranges are vast! Condors can easily cover hundreds of miles in a day (as the regular occurrence of birds in southern Utah attests). If the condor reintroduction project meets its objectives (150 birds in southern California and 150 birds in northern Arizona), condors may be expected to wander

California Condor, underside. Juvenile gliding **(A)**. Adult soaring **(B)**. Amazingly large and steady in flight; many-fingered wingtips obvious at all angles, as long primaries flex and twist. Wings broad and heavy, pushed forward when soaring, can appear very pointed when gliding or when viewed from side; tail short and square with sharp corners. Whitish underwing coverts and silvery upperwing coverts distinctive.

California Condor, upperside. Adult soaring.

and possibly repopulate areas where they are not yet found.

Until then, look for condors near reintroduction sites. These include Sespe Condor Sanctuary and Lion Canyon in Los Padres National Forest. Other sites include Castle Crags Wilderness and Ventana Wilderness near Big Sur in California. In Arizona, try the Grand Canyon region, specifically the Vermilion Cliffs and Marble Canyon, areas where condors were known to survive in fairly recent times. Perhaps the most productive condor-watching location lies in the vicinity of El Tovar Hotel, on the South Rim, where the abundance of large mammals (tourists, perhaps regarded as prey) attracts large numbers of condors in summer.

Check hotlines and nature centers; talk to other birders for the latest information, as the birds can and do move around. Respect boundary signs and the birds. The birds are best served by limiting human contact; besides, condors are so big that you can enjoy them a mile away (and identify them much farther).

California Condor is huge — the largest bird in North America and the second-largest bird in the world (second only to its close relative, the Andean Condor). Wingspans to nearly 10 feet and weights of more than 30 pounds have been recorded.

California Condor soaring head-on; wings held nearly flat while soaring, with brushlike "fingers" at wingtips.

While California Condor greatly resembles Old World vultures (such as the Griffon Vulture), it is related to the other New World vultures — Turkey, Black, and King. As such, the California Condor is more allied to storks than to the other raptors included in this book. Similarities between the condor and Old World vultures, relatives of true raptors, are a consequence of convergent evolution.

Were California Condors widespread, they would have figured in the chapter treating eagles and vultures, because they are, without question, *big* black birds — *far* larger than Bald Eagle or Golden Eagle. And *black* — like the black of Black Vulture (not the dark brown of Turkey Vulture or Golden Eagle). Juveniles, like the birds most commonly found near release sites, are blackest of all.

Adults have a bare, orange head, while the head of juveniles is black. The orange head of adults, while obvious and bright, is not always apparent under field conditions. Adults show a bright white elongated triangular patch on the underwing coverts. This patch may be mottled at the base of the primaries. Juveniles show mottled, dusky underwing patches, which are gray at a distance and may be difficult to see. Condors take five or more years to reach adult plumage, and immatures of intermediate age show considerable variation on the underwings and head.

Above, adults show a white border along the inner edge of the wing coverts and a silvery wash over the secondaries. These features, too, can be hard to see at a distance.

IN FLIGHT. The California Condor appears massive. This sense is projected by actual size, by the slow-motion quality of the bird's flight, and often by direct comparison of condor with other soaring birds that it, of course, dwarfs (including eagles). Case in point: In direct comparison, it takes 8 seconds for a soaring Red-tailed Hawk to complete a circle; 13 seconds for a Golden Eagle; 16 seconds for a condor.

Wings are long, broad, and heavy, and pushed forward in a soar. The outer primaries are deeply splayed. Extending fingerlike from the bird's wingtips, these great airfoils are obvious at almost any angle and at great distances.

The trailing edge of the wing has a distinct S-shape — the structural product of bulging secondaries, shortened inner primaries, and wide outer primaries that curl back. In a glide, with primaries closed and pulled back, the wings appear extremely pointed — even triangular. Overall, the wings of California Condor have a remarkable protean quality, their shape conforming and contouring to accommodate conditions and condor ambitions.

The head seems comically small for such an immense bird, but it droops conspicuously. The tail is of medium length, proportionally shorter than Turkey Vulture's but considerably longer than Black Vulture's. The condor actively uses the tail, spreading, closing, and angling it for balance. At a distance, the tail appears squared off and sharp at the corners.

In a glide the condor holds its wings level or with a slight, curving dihedral. The slotted primaries curl up at the tips (more so than the outer primaries of eagles), and the angle may be so acute that the feathers seem cocked rather than curved. Occasionally, California Condor flies with a modified dihedral, with the inner wing raised but the outer wing flat. Like Black Vultures, condors sometimes soar with feet down and, like Turkey Vultures, they execute a "wing dip" maneuver in which both wings seem to wilt momentarily, then return to a stiff-winged glide.

When a soaring bird begins a glide, it takes one or two deep wingbeats, bending the wings so far below the body that the tips seem almost to touch. Once gliding, the bird covers great distances on set wings without flapping and covers ground *fast*. Condors have been known to travel more than 40 miles in an hour, including the time spent soaring to gain altitude.

Condors are amazing fliers. They forage great distances from nest ledges and roosts by making extensive use of thermals. They convert altitude into distance in fast flat glides but seem not to lose altitude in the process (suspending the laws of physics). At other times, they appear to fly in slow motion, covering little ground, even remaining in place. When they are soaring, turns can be on flat wings — that is, they appear to bank very little when turning.

When condors do flap, they appear ponderous, lumbering. The flap is very slow, yet smooth and deep, and almost entirely below the horizontal. The wing appears to bend along the entire length. When soaring and gliding, condors flap little. Yet, around roosts and carcasses, they are surprisingly animate, energetic in their flapping, and (even more surprisingly) agile, engaging in tail chases low through scattered trees.

Condors are gregarious and social; they fly in obvious squadrons. In formation, the birds frequently fly parallel to one another, much like Black Vultures, but also use the "line astern" groupings of Turkey Vultures. When soaring in a thermal, individuals will often turn in unison, like American White Pelicans, but may also circle randomly, like Broad-winged Hawks.

There is little chance of confusing a condor with one of the two smaller vulture species. Condors dwarf Turkey and Black vultures, and the differences only begin here. Turkey Vultures wobble in flight, condors are steady. Black Vultures, miniature condors, have structural kinship with California Condors — but, as yet, the two species do not overlap in range.

Confusing California Condor and eagles is more likely. Immature Bald Eagles have an underwing pattern that is similar to the condor's and a ponderous flap (except Bald Eagle raises the wing mostly above the body and condors are masters of the downstroke). Eagles and condors also hold their wings level or with a slight dihedral, and both birds also adjust their flights by flexing their wings at the wrist.

Once again, it comes down to size. Condors, with their nine-foot wingspan, dwarf the six-and-a-half-foot wingspan of Golden Eagle. Seen together in a thermal, condors outsize Golden Eagles the way Golden Eagles dwarf Red-tailed Hawks. With the condor's distinctive, flat-winged profile and steady, sky-crossing glides, there is little danger of mistaking it for anything except an airplane.

One last clue, a card rarely played in the hawk-watching arena, relates to sound.

No, California Condors don't call, but the swoosh of their wingbeats can be heard for hundreds of yards. The hiss or whistle of wind passing through the slotted primaries also carries and may, under ideal conditions, be heard a hundred yards away.

But if all else fails and you are still uncertain about the identification of a probable condor, you can confirm this identification by noting the large colored plastic wing markers affixed to the leading edge of each wing. These markers are used by condor biologists to track the movements of individual birds (some birds have visible radio antennae, too). They may not be natural field marks, and they may detract from the aesthetic appreciation of the encounter. But they are a poignant reminder of how close we came to losing North America's greatest soaring bird. If this is the cost of seeing condors in the wild, it's way underpriced.

1. A soaring adult California Condor. The bare orange head and bright white underwing linings are hard to miss. Arizona, June CS

2. An adult condor banking. The birds can appear remarkably short tailed and pointy winged at times. Arizona, June CS

3. An adult condor in a full soar, with a full crop and full bling. The crop is the pink bulge about midchest. The bling, the wing tags worn by this intensely studied and managed species, look like price tags set on either wing. Arizona, June CS

4. Two adult condors in a fast glide. The wings are turned severely back, but tails remain broad. You would not be the first to be reminded of the shape of Black Vulture. Arizona, June CS

5. A juvenile California Condor with a Common Raven escort. Juveniles show much darker underwings (and a gray head, not orange). Note, particularly, the very deeply slotted outer primaries. Arizona, June CS

6. A juvenile condor. In direct sunlight, the more limited white on the underwing lining can be very apparent, but note the difference in the shaded wing. Arizona, June CS

7. A two-year-old condor (showing extensive wing molt) riding an updraft. Note the humpbacked appearance, the modest dihedral, and the deeply slotted primaries. This is no Golden Eagle coming your way! Habituated to people, birds like this can be enjoyed by visitors to the Grand Canyon. Arizona, June CS

8. An adult California Condor soaring in the Grand Canyon, from above. The eagle-dwarfing size of this species is something that cannot be projected in an isolated image, especially when photographed against the Grand Canyon (also see p. 298). Arizona, June CS

Other Birds That Soar

This book began on an optimistic note, with the assurance that learning to identify distant birds of prey is a skill that anyone can master. Perhaps it's only proper that it should end on a cautionary note.

In your focus on birds of prey, never lose sight of the fact that raptors are not the only birds that soar. Other species habitually soar (and glide), many with as much finesse as any bird of prey. At times, and under some conditions, even birds that are not regarded as soaring species (like some shorebirds) will get in a thermal and do a turn or two. Be warned. Be ready.

GULLS

Gulls are often mistaken for birds of prey (particularly Osprey and White-tailed Kite). Gulls are large. They are, in the juvenile stages of many gull species, brown or brown and white like many raptor species. They soar and glide beautifully and, in a kettle, circle independently (like hawks), not in synchronized formation (like cranes or white pelicans).

Shape is perhaps the most distinguishing characteristic. Gulls have long, slim, and very pointed wings, traits that easily distinguish them from most of the large soaring hawks.

Unlike most birds of prey, gulls also have a large prominent head that projects ahead of the wing as much as the tail projects behind. Also, a down-crooked gull-like wing is not characteristic of most soaring raptors. Among birds of prey, only Osprey, Swallow-tailed Kite, and Snail Kite might be accused of showing this gull-winged configuration to the degree evidenced on a gull species.

One important thing to bear in mind: When raptors soar, the outer primaries splay out, like fingers on a hand. The wingtips of gulls just stay pointy. You can't distinguish individual feathers on a gull's wing. You can on a hawk.

The wingbeat of gulls is fairly slow, languid, and shallow. Most birds of prey put more oomph into it.

TOP LEFT: American White Pelicans. Circling in unison, a flock of pelicans soar on tropical thermals. It's a kettle, but they sure aren't hawks. Veracruz, October CS

BOTTOM LEFT: A Wood Stork. Perhaps related to New World Vultures, it is a big black-and-white bird, not a big black bird, but it soars much like them (although flocks usually turn in unison). South Carolina, September CS

BOTTOM RIGHT: An Anhinga soaring. At great height, an Anhinga, when either soaring or gliding, can look deceptively like a falcon or, particularly, a kite. This is because, owing to distance, the snaky neck and head can disappear, leaving a shape much like that of a long-tailed hawk. Florida, March CS

RAVENS (AND CROWS)

Functional raptors, these large corvids are *very* hawklike in flight and may easily be mistaken for a distant eagle or dark-morph buteo. Once again, shape is very helpful. If you are fairly close, the corvid's straight bill will end confusion. The wedge-shaped tail of ravens is also unlike the tail of most raptors.

Perhaps the most distinguishing characteristic is a soaring raven's tendency to angle the very tip of its wing back. Raptors *glide* with wingtips back, but soar with the outer flight feathers straight out, not curled back. This may seem like a subtle trait, but it is surprisingly evident and may be used to distinguish ravens from raptors at great distances.

Crows are sometimes mistaken for raptors but are smaller (in most cases much smaller) than any dark-morph buteo, and, while they can soar and glide, they are not in the class of a hawk or raven. Sooner or later, a crow is going to flap, and the steady, rowing wingbeat will help distinguish it.

ANHINGAS, CORMORANTS, PELICANS, STORKS, CRANES (AND SOME HERONS)

All of these species soar and migrate in large flocks, as do several raptor species. Individually, and at anything approaching a reasonable distance, each of these species is easily distinguished from a soaring bird of prey by shape alone. But at distances measured in miles, birds are little more than specks. Shape, at these distances, becomes meaningless.

One characteristic that all these species share that easily distinguishes them from kettling hawks like Broad-winged or Swainson's is their penchant to soar in synchronized formation, all turning in the same direction. Kettling raptors soar in an unsynchronized maelstrom, swirling in all directions, every hawk for itself.

OTHER POSSIBLE CANDIDATES FOR CONFUSION

Many owl species are also migratory, and one species, the pale Short-eared Owl, is occasionally seen in migration. Its mothlike flight may recall a harrier, but it is easily distinguished by shape. Harriers are long and lanky; Short-eared Owls look like beer kegs on wings.

Swallows, particularly the Purple Martin, glide well and have very falconlike shapes. But their flight is quick, erratic, and darting. Falcons are direct.

Pigeons are also very falconlike — in fact, they are easily mistaken for Merlins. But pigeons often fly in tight flocks, something that no Merlin has ever dreamed of doing. Pigeons also twist their bodies in flight, angling down on one side, then the other; Merlins keep an even keel. And while the wingbeats of pigeons can be steady, they frequently insert one or two halting, arrhythmic wingbeats into a series. Merlins have two speeds, fast and faster — but the cadence of the wingbeats is steady. Merlins may accelerate, but they never miss a beat.

RAPTOR CONSERVATION

To be interested in birds of prey is to be concerned about their welfare, and to this objective the authors of this book have dedicated large portions of our lives.

As someone who shares our interest and concern, it is incumbent upon you to share your knowledge and your passion with people who, as yet, lack your insight.

A migrant adult Peregrine Falcon, flapping as it begins a turn, pausing to check out a hawk watcher before it aims for the horizon anew. Cape May, November TJ

Birds of prey can be relied upon to inspire us. They always have. It's up to us to ensure that they always do.

Someone, sometime, ushered you into the ranks of hawk watchers. The way to say thanks is to pass their gift on. The son, granddaughter, friend, or colleague whom you take to your favorite hawk watch will have you to thank for the rest of their days.

It's not only birds of prey that need your support. Organizations dedicated to their welfare do, too. Your membership is what gives them the means and the latitude to do their important work. In the end, the support you give comes back to you. Every spring. Every fall. Every time you raise your binoculars and are treated to the vision of hawks in flight.

Bibliography

Bechard, M. J., and J. K. Schmutz. "Ferruginous Hawk." In *The Birds of North America,* no. 172. Edited by A. Poole and F. Gill. Philadelphia, PA: Academy of Natural Sciences, and Washington, DC: American Ornithologists' Union, 1995.

Bechard, M. J., and T. R. Swem. "Rough-legged Hawk." In *The Birds of North America,* no. 641. Edited by A. Poole and F. Gill. Philadelphia, PA: Birds of North America, Inc., 2002.

Beebe, Frank L. *Field Studies of the Falconiformes of British Columbia.* Victoria, BC: British Columbia Provincial Museum, 1974.

Bibles, B. D., R. L. Glinski, and R. R. Johnson. "Gray Hawk." In *The Birds of North America,* no. 652. Edited by A. Poole and F. Gill. Philadelphia, PA: Birds of North America, Inc., 2002.

Bildstein, K. L., and K. Meyer. "Sharp-shinned Hawk." In *The Birds of North America,* no. 482. Edited by A. Poole and F. Gill. Philadelphia, PA: Birds of North America, Inc., 2000.

Brett, James, and Alex Nagy. *Feathers in the Wind.* Kempton, PA: Hawk Mountain Sanctuary Association, 1973.

Broun, Maurice. *Hawks Aloft.* Kutztown, PA: Kutztown Publishing, 1948.

Brown, Leslie. *Birds of Prey: Their Biology and Ecology.* New York: A & W Publishers, 1977.

Brown, Leslie, and Dean Amadon. *Eagles, Hawk, and Falcons of the World.* New York: McGraw-Hill, 1968.

Buckley, N. J. "Black Vulture." In *The Birds of North America,* no. 411. Edited by A. Poole and F. Gill. Philadelphia, PA: Birds of North America, Inc., 1999.

Buehler, D. A. "Bald Eagle." In *The Birds of North America,* no. 506. Edited by A. Poole and F. Gill. Philadelphia, PA: Birds of North America, Inc., 2000.

Cade, Tom J. *The Falcons of the World.* London: William Collins, 1982.

Clark, William S. "Extreme Variation in the Tails of Adult Harlan's Hawks." *Birding* 41: 30–36.

_____. "Field Identification of Accipiters in North America." *Birding* 16, no. 6 (December 1984): 251–263.

Clark, William S., and Brian K. Wheeler. *A Field Guide to Hawks of North America.* 2d ed. Boston: Houghton Mifflin, 2001.

Clum, N. J., and T. J. Cade. "Gyrfalcon." In *The Birds of North America,* no. 114. Edited by A. Poole and F. Gill. Philadelphia, PA: Academy of Natural Sciences, and Washington, DC: American Ornithologists' Union, 1994.

Crocoll, S. T. "Red-shouldered Hawk." In *The Birds of North America,* no. 107. Edited by A. Poole and F. Gill. Philadelphia, PA: Academy of Natural Sciences, and Washington, DC: American Ornithologists' Union, 1994.

Dunk, J. R. "White-tailed Kite." In *The Birds of North America,* no. 178. Edited by A. Poole and F. Gill. Philadelphia, PA: Academy of Natural Sciences, and Washington, DC: American Ornithologists' Union, 1995.

Dunne, Pete. "How to Tell a Hawk from a Gull: The Road That Led from the Shotgun to the Subjective." *Newsletter of the Hawk Migration Association of North America* 10, no. 1 (February 1985): 8–10.

______. *Pete Dunne's Essential Field Guide Companion.* Boston: Houghton Mifflin, 2006.

Dunne, Pete, Debbie Keller, and Rene Kochenberger. *Hawk Watch: A Guide for Beginners.* Cape May, NJ: New Jersey Audubon Society, 1984.

Dunne, Pete, David Sibley, Clay Sutton, and Fred Hamer. "The Falcon That Isn't: The Mississippi Kite." *Newsletter of the Hawk Migration Association of North America* 8, no. 1 (February 1983): 22–24.

______. "Field Identification of Broad-winged and Red-shouldered Hawk." *Newsletter of the Hawk Migration Association of North America* 7, no. 1 (February 1982): 8–9.

______. "Wing on at the Limit of Conjecture." *Newsletter of the Hawk Migration Association of North America* 7, no. 2 (August 1982): 9–10.

Dunne, Pete, and Clay Sutton. "Population Trends in Coastal Raptor Migrants over Ten Years of Cape May Point Autumn Counts." *Records of New Jersey Birds* 12, no. 3 (Autumn 1986): 39–43.

Dunne, Pete, Clay Sutton, and David Sibley. "Zone-tailed Hawk." *Birding* 32 (June 2000): 234–241.

England, A. S., M. J. Bechard, and C. S. Houston. "Swainson's Hawk." In *The Birds of North America,* no 265. Edited by A. Poole and F. Gill. Philadelphia, PA: Academy of Natural Sciences, and Washington, DC: American Ornithologists' Union, 1997.

Farquhar, C. C. "White-tailed Hawk." In *The Birds of North America,* no. 30. Edited by A. Poole, P. Stettenheim, and F. Gill. Philadelphia, PA: Academy of Natural Sciences, and Washington, DC: American Ornithologists' Union, 1992.

Ferguson-Lees, James, and David A. Christie. *Raptors of the World.* Boston: Houghton Mifflin, 2001.

Goodrich, L. J., S. C. Crocoll, and S. E. Senner. "Broad-winged Hawk." In *The Birds of North America,* no. 218. Edited by A. Poole and F. Gill. Philadelphia, PA: Academy of Natural Sciences, and Washington, DC: American Ornithologists' Union, 1996.

Harwood, Michael, ed. *Proceedings of the Hawk Migration Conference, IV.* Rochester, NY: Hawk Migration Association of North America, 1983.

Heintzelman, Donald S. *Autumn Hawk Flights: The Migrations in Eastern North America.* New Brunswick, NJ: Rutgers University Press, 1975.

______. *A Guide to Hawk Watching in North America,* 2d ed. Guilford, CT: Globe Pequot Press, 2004.

Johnsgard, Paul A. *Hawks, Eagles, and Falcons of North America.* Washington DC: Smithsonian Institution, 1990.

Johnson, R. R., R. L. Glinski, and S. W. Matteson. "Zone-tailed Hawk." In *The Birds of North America,* no. 529. Edited by A. Poole and F. Gill. Philadelphia, PA: Birds of North America, Inc., 2000.

Journal of the Hawk Migration Association of North America 2, no. 1 (December 1980).

Julian, Paul R. "Harlan's Hawk: A Challenging Taxonomic and Field Problem." *Colorado Field Ornithology,* no. 1 (Winter 1967): 1–6.

Keddy-Hector, D. P. "Aplomado Falcon." In *The Birds of North America,* no. 549. Edited by A. Poole and F. Gill. Philadelphia, PA: Birds of North America, Inc., 2000.

Kirk, D. A., and M. J. Mossman. "Turkey Vulture." In *The Birds of North America,* no. 339. Edited by A. Poole and F. Gill. Philadelphia, PA: Birds of North America, Inc., 1998.

Kochert, M. N., K. Steenhof, C. L. McIntyre, and E. H. Craig. "Golden Eagle." In *The Birds of North America,* no. 684. Edited by A. Poole and F. Gill. Philadelphia, PA: Birds of North America, Inc., 2002.

Langley, Lynn. "Swallowtails of Francis Marion." *South Carolina Wildlife* 31, no. 5 (September–October 1984): 6–10.

Liguori, Jerry, "Dark Red-tailed Hawks." *Birding* 36: 500–506.

______. *Hawks at a Distance.* Princeton, NJ: Princeton University Press, 2011.

______. *Hawks from Every Angle.* Princeton, NJ: Princeton University Press, 2005.

______. "Pitfalls of Classifying Light Morph Red-tailed Hawks to Subspecies." *Birding* 33: 436–446.

Liguori, Jerry and B. L. Sullivan. "Comparison of Harlan's with Western and Eastern Red-tailed Hawks." *Birding* 42: 30–37.

______. "A Study of Krider's Red-tailed Hawk." *Birding* 42: 38–45.

Lish, James W., and William G. Voelker. "Field Identification Aspects of Some Red-tailed Hawk Subspecies." *American Birds* 40, no. 2 (Summer 1986): 197–202.

MacWhirter, R. B., and K. L. Bildstein. "Northern Harrier." In *The Birds of North America,* no. 210. Edited by A. Poole and F. Gill. Philadelphia, PA: Academy of Natural Sciences, and Washington, DC: American Ornithologists' Union, 1996.

Meyer, K. D. "Swallow-tailed Kite." In *The Birds of North America,* no. 138. Edited by A. Poole and F. Gill. Philadelphia, PA: Academy of Natural Sciences, and Washington, DC: American Ornithologists' Union, 1995.

Miller, K. E., and K. D. Meyer. "Short-tailed Hawk." In *The Birds of North America,* no. 674. Edited by A. Poole and F. Gill. Philadelphia, PA: Birds of North America, Inc., 2002.

Mindell, David P. "Plumage Variation and Winter Range of Harlan's Hawk." *American Birds* 39, no. 2 (Summer 1981): 127–133.

Morrison, J. L. "Crested Caracara." In *The Birds of North America,* no. 249. Edited by A. Poole and F. Gill. Philadelphia, PA: Academy of Natural Sciences, and Washington, DC: American Ornithologists' Union, 1996.

Mullarney, Killian, Lars Swensson, Dan Zetterstrom, and Peter J. Grant. *Birds of Europe.* Princeton, NJ: Princeton University Press, 1999.

Muller, Helmut C., Daniel D. Berger, and George Allex. "The Identification of North American Accipiters." *American Birds* 33, no. 3 (May 1979): 236–240.

Newsletter of the Hawk Migration Association of North America. Seasonal reports, 1977–1985.

Newton, Ian. *Population Ecology of Raptors.* Vermillion, SD: Buteo Books, 1979.

Parker, J. W. "Mississippi Kite." In *The Birds of North America,* no. 402. Edited by A. Poole and F. Gill. Philadelphia, PA: Birds of North America, Inc., 1999.

Peterson, Roger Tory. *Peterson Field Guide to Birds of Eastern and Central North America.* 6th ed. Boston: Houghton Mifflin Harcourt, 2010.

______. *Peterson Field Guide to Birds of North America.* Boston: Houghton Mifflin Harcourt, 2008.

______. *Peterson Field Guide to Birds of Western North America.* 4th ed. Boston: Houghton Mifflin Harcourt, 2010.

Poole, A. F., R. O. Bierregaard, and M. S. Martell. "Osprey." In *The Birds of North America,* no. 683. Edited by A. Poole and F. Gill. Philadelphia, PA: Birds of North America, Inc., 2002.

Porter, R. F., Ian Willis, Steen Christensen, and Bent Pors Nielsen. *Flight Identification of European Raptors.* Calton, England: T. & A. D. Poyser, 1981.

Preston, C. R., and R. D. Beane. "Red-tailed Hawk." In *The Birds of North America,* no. 52. Edited by A. Poole and F. Gill. Philadelphia, PA: Academy of Natural Sciences, and Washington, DC: American Ornithologists' Union, 1993.

Radcliffe, Derek. *The Peregrine Falcon.* Vermillion, SD: Buteo Books, 1980.

Raynord, Edward J., Ted T. Cable, and David Wiggins. "Common Black-Hawk: A Probable Kansas Record and Possible Recent Range Expansion." *Kansas Ornithological Society Bulletin.* December 2012.

Robbins, Chandler S., Bertel Bruun, and Herbert S. Zim. *Birds of North America.* New York: Golden Press, 1966.

Rosenfeld, R. N., and J. Bielefeldt. "Cooper's Hawk." In *The Birds of North America,* no. 75. Edited by A. Poole and F. Gill. Philadelphia, PA: Academy of Natural Sciences, and Washington, DC: American Ornithologists' Union, 1993.

Schnell, J. H. "Common Black-Hawk." In *The Birds of North America,* no. 122. Edited by A. Poole and F. Gill. Philadelphia, PA: Academy of Natural Sciences, and Washington, DC: American Ornithologists' Union, 1994.

Scott, Shirley L., ed. *Field Guide to the Birds of North America.* Washington, DC: National Geographic Society, 1983.

Sibley, David Allen. *The Sibley Guide to Birds.* New York: Alfred A. Knopf, 2000.

Smallwood, J. A., and D. M. Bird. "American Kestrel." In *The Birds of North America,* no. 602. Edited by A. Poole and F. Gill. Philadelphia, PA: Birds of North America, Inc., 2002.

Sodhi, N. S., L. W. Oliphant, P. C. James, and I. G. Warkentin. "Merlin." In *The Birds of North America,* no. 44. Edited by A. Poole and F. Gill. Philadelphia, PA: Academy of Natural Sciences, and Washington, DC: American Ornithologists' Union, 1993.

Squires, J. R., and R. T. Reynolds. "Northern Goshawk." In *The Birds of North America,* no. 298. Edited by A. Poole and F. Gill. Philadelphia, PA: Academy of Natural Sciences, and Washington, DC: American Ornithologists' Union, 1997.

Steenhoff, K. "Prairie Falcon." In *The Birds of North America,* no. 346. Edited by A. Poole and F. Gill. Philadelphia, PA: Birds of North America, Inc., 1998.

Sutton, Clay C. "Identification Review: Broad-winged Hawks in Spring—Red-shouldered Imposters." *Birding* 34 (April 2002): 176–180.

Sutton, Clay C., and Patricia Taylor Sutton. *How to Spot Hawks and Eagles.* Boston: Houghton Mifflin, 1996.

______. "Rio de Rapaces." In *Pronatura, La Conservation de la Natualeza en Mexico* (Pronatura, 2000).

______. "The River of Raptors: Exploring and Enjoying Pronatura Veracruz's Raptor Conservation Project." *Birding* 31 (June 2000): 229–236.

______. "The Spring Hawk Migration at Cape May, New Jersey." *Cassinia* 60 (1983): 5–18.

Sutton, Clay C., and Richard Walton. *Birds of Prey of North America.* National Audubon Society Pocket Guide Series. New York: Chanticleer Press, Knopf, 1994.

Sykes, P. W., Jr., J. A. Rodgers Jr., and R. E. Bennetts. "Snail Kite." In *The Birds of North America,* no. 171. Edited by A. Poole and F. Gill. Philadelphia, PA: Academy of Natural Sciences, and Washington, DC: American Ornithologists' Union, 1995.

Watson, Donald. *The Hen Harrier.* Berkhamsted, Hertfordshire, England: T. & A. D. Poyser, 1977.

Wheeler, Brian K. *Raptors of Eastern North America.* Princeton, NJ: Princeton University Press, 2003.

______. *Raptors of Western North America.* Princeton, NJ: Princeton University Press, 2003.

Wheeler, Brian K., and William S. Clark. *A Photographic Guide to North American Raptors.* London: Academic Press, 1995.

White, Clayton M., and Tom J. Cade. "Cliff Nesting Raptors and Ravens Along the Colville River in Arctic Alaska." In *The Living Bird.* Ithaca, NY: Cornell Laboratory of Ornithology, 1971.

White, C. M., N. J. Clum, T. J. Cade, and W. G. Hunt. "Peregrine Falcon." In *The Birds of North America,* no. 660. Edited by A. Poole and F. Gill. Philadelphia, PA: Birds of North America, Inc., 2002.

Index

Page references in italics refer to text graphics.

LEFT: One of the ultimate rewards of hawk watching: a gray morph Gyrfalcon patrolling the high Arctic tundra. Alaska, June TS

Photographer Key

WC	William S. Clark
MC	Merrill Cottrell
LD	Linda Dunne
DF	Don Freiday
NH	Ned Harris
TJ	Tom Johnson
KK	Kevin T. Karlson
TL	Tony Leukering
JL	Jerry Liguori
JLISH	Jim Lish
MOB	Michael O'Brien
LO	Luke Ormond
CS	Clay Sutton
TS	Ted Swem

Short-tailed
Hawk
Gray
Hawk
Aplomado
Falcon
Crested
Caracara
White-tailed
Hawk
Harris'
Hawk
Snail
Kite
Hook-billed
Kite
California
Condor
Common
Black-Hawk
Zone-tailed
Hawk